Farmers, Gene Banks and Crop Breeding:

Economic Analyses of Diversity in

Wheat, Maize, and Rice

NATURAL RESOURCE MANAGEMENT AND POLICY

Editors:

Ariel Dinar
Agricultural and Natural Resources Dept.
The World Bank
1818 H Street, NW
Washington, DC 20433

David Zilberman
Dept. of Agricultural and
Resource Economics
Univ. of California, Berkeley
Berkeley, CA 94720

EDITORIAL STATEMENT

There is a growing awareness to the role that natural resources such as water, land, forests and environmental amenities play in our lives. There are many competing uses for natural resources, and society is challenged to manage them for improving social well being . Furthermore, there may be dire consequences to natural resources mismanagement. Renewable resources such as water, land and the environment are linked, and decisions made with regard to one may affect the others. Policy and management of natural resources now require interdisciplinary approach including natural and social sciences to correctly address our society preferences.

This series provides a collection of works containing most recent findings on economics, management and policy of renewable biological resources such as water, land, crop protection, sustainable agriculture, technology, and environmental health. It incorporates modern thinking and techniques of economics and management. Books in this series will incorporate knowledge and models of natural phenomena with economics and managerial decision frameworks to assess alternative options for managing natural resources and environment.

Concerns about genetic diversity of agricultural crops grow as traditional varieties are replaced by modern varieties, and as biotechnology threatens genetic diversity to further deteriorate. This book is the first comprehensive effort to bring together the state of the art of the economic analysis of genetic diversity. The book develops new applications in economics using methods such as sequential search models, probability models for gene distributions, and household models. It offers first empirical results of costs and benefits of genetic diversity.

The Series Editors

Recently Published Books in the Series

Antle, John, Capalbo, Susan and Crissman, Charles:
Economic, Environmental, and Health Tradeoffs in Agriculture:
Pesticides and the Sustainability of Andean Potato Production
Spulber, Nicolas and Sabbaghi, Asghar:
Economics of Water Resources: From Regulation to Privatization
Second Edition
Bauer, Carl J.:
Against the Current: Privatization, Water Markets, and the State in Chile
Easter, K. William, Rosegrant, Mark W., and Dinar, Ariel:
Markets for Water: Potential and Performance

Farmers, Gene Banks and Crop Breeding:

Economic Analyses of Diversity in
Wheat, Maize, and Rice

EDITED BY
Melinda Smale

International Maize and Wheat Improvement Center
Mexico

KLUWER ACADEMIC PUBLISHERS
Boston / Dordrecht / London

Distributors for North, Central and South America:
Kluwer Academic Publishers
101 Philip Drive
Assinippi Park
Norwell, Massachusetts 02061 USA
Telephone (781) 871-6600
Fax (781) 871-6528
E-Mail <kluwer@wkap.com>

Distributors for all other countries:
Kluwer Academic Publishers Group
Distribution Centre
Post Office Box 322
3300 AH Dordrecht, THE NETHERLANDS
Telephone 31 78 6392 392
Fax 31 78 6546 474
E-Mail <services@wkap.nl>

 Electronic Services <http://www.wkap.nl>

Library of Congress Cataloging-in-Publication Data

A C.I.P. Catalogue record for this book is available
from the Library of Congress.

ISBN 978-1-4419-5068-0

Printed on acid-free paper.

Printed in the United States of America

CONTENTS

FIGURES

TABLES

CONTRIBUTING AUTHORS

Girlie Abrigo
Philippine Rice Research Institute (PhilRice)
Maligaya, Muñoz, Philippines

Alfonso Aguirre
Human Ecologist
Economics Program
International Maize and Wheat Improvement Center
  (CIMMYT)
Mexico, D.F., Mexico

Mauricio R. Bellon
Human Ecologist
Economics Program
International Maize and Wheat Improvement Center
  (CIMMYT)
Mexico, D.F., Mexico

Stephen B. Brush
Professor
Department of Human and Community Development
University of California-Davis
Davis, California, USA

Marlon Calibo
Genetic Resources Center
International Rice Research Institute (IRRI)
Manila, Philippines

David A. Cleveland
Associate Professor
Department of Anthropology
University of California-Santa Barbara
Santa Barbara, California, USA

Dennis Erasga
Genetic Resources Center
International Rice Research Insitute (IRRI)
Manila, Philippines

Robert E. Evenson
Professor
Economic Growth Center
Yale University
New Haven, Conneticut, USA

Sergio R. Francisco
Philippine Rice Research Institute (PhilRice)
Maligaya, Muñoz, Philippines

Douglas Gollin
Assistant Professor of Economics
Williams College
Fernald House
Williamstown, Massachusetts, USA
Affiliate Economist, CIMMYT, Mexico

Daniel Grimanelli
Geneticist
Applied Biotechnology Center
International Maize and Wheat Improvement Center
  (CIMMYT)
Mexico, D.F., Mexico

Jason Hartell
Research Associate
Department of Agricultural Economics
Katholieke Universiteit Leuven
Heverlee, Belgium

Amir Heiman
Lecturer
Department of Agricultural and Resource Economics
Hebrew University of Jerusalem
Israel

Paul W. Heisey
Economist
Economics Program
International Maize and Wheat Improvement Center
  (CIMMYT)
Mexico, D.F., Mexico

Stéphane Lemarié
Economist
Chargé de Recherche
Institut National de la Recherche Agronomique
  Sociologie et Économie de la Recherche-
  Développement (INRA-SERD)
Université Pierre Mendès-France
Grenoble, France

Genoveva C. Loresto
Philippine Rice Research Institute (PhilRice)
Maligaya, Muñoz, Philippines

Dominique Louette
Agronomist
University of Guadalajara
Instituto Manantlan de Ecología y Conservación de
  la Biodiversidad (IMECBIO)
Guadalajara, Jalisco, Mexico

xii

Erika C. H. Meng
Economist
Economics Program
International Maize and Wheat Improvement Center
    (CIMMYT)
Mexico, D.F., Mexico

Michael L. Morris
Economist
Economics Program
International Maize and Wheat Improvement Center
    (CIMMYT)
Mexico, D.F., Mexico

Phil G. Pardey
Economist
International Food Policy Research Institute (IFPRI)
Washington, DC, USA

Jean-Louis Pham
Geneticist
Genetic Resources Center
International Rice Research Institute (IRRI)
Manila, Philippines
Seconded to IRRI by ORSTOM
(Institut Français de Recherche Scientifique pour le
    Développement en Coopération)
Paris, France

Hugo Perales R.
Department of Applied Behavioral Sciences
University of California-Davis
Davis, California, USA

Prabhu L. Pingali
Director
Economics Program
International Maize and Wheat Improvement Center
    (CIMMYT)
Mexico, D.F., Mexico

Calvin O. Qualset
Director
Genetic Resources Conservation Program
University of California-Davis
Davis, California, USA

Sheila Quilloy
Genetic Resources Center
International Rice Research Institute (IRRI)
Manila, Philippines

Elizabeth B. Rice
Department of Soil, Crop and Atmospheric Sciences
Cornell University
Ithaca, New York, USA

Scott Rozelle
Associate Professor
Department of Agricultural and Resource Economics
University of California-Davis
Davis, California, USA

Paul Sanchez
Philippine Rice Research Institute (PhilRice)
Maligaya, Muñoz, Philippines

Leocadio S. Sebastian
Philippine Rice Research Institute (PhilRice)
Maligaya, Muñoz, Philippines

Ben Senauer
Director of the Center for International Food and
    Agricultural Policy
Professor, Department of Applied Economics
University of Minnesota
St. Paul, Minnesota, USA

Ravi P. Singh
Geneticist-Pathologist
Wheat Program
International Maize and Wheat Improvement Center
    (CIMMYT)
Mexico, D.F., Mexico

Bent Skovmand
Head, Wheat Genetic Resources Center
International Maize and Wheat Improvement Center
    (CIMMYT)
Mexico, D.F., Mexico

Melinda Smale
Economist
Economics Program
International Maize and Wheat Improvement Center
    (CIMMYT)
Mexico, D.F., Mexico

Daniela Soleri
Arid Lands Resource Sciences
Office of Arid Land Studies
University of Arizona
Tucson, Arizona, USA

Suketoshi Taba
Head, Maize Genetic Resources Center
International Maize and Wheat Improvement Center
    (CIMMYT)
Mexico, D.F., Mexico

J. Edward Taylor
Professor
Department of Agricultural and Resource Economics
University of California-Davis
Davis, California, USA

Greg Traxler
Associate Professor
Department of Economics
Auburn University
Auburn, Alabama, USA
Affiliate Economist, CIMMYT, Mexico

Eric Van Dusen
Department of Agricultural and Resource Economics
University of California-Davis
Davis, California, USA

David Widawsky
Associate Economist
Social Sciences Division
International Rice Research Institute (IRRI)
Manila, Philippines

Brian D. Wright
Professor
Department of Agriculture and Resource Economics
University of California-Berkeley
Berkeley, California, USA

Cherisa Yarkin
Associate Director for Economic Analysis
Critical Linkage Project
University of California Systemwide Biotechnology
    Research and Education Program

David Zilberman
Professor
Department of Agricultural and Resource Economics
Member
Giannini Foundation of Agricultural Economics
University of California-Berkeley
Berkeley, California, USA

PREFACE

This book has its origins in the research community's concern for crop genetic resources and genetic diversity. At the International Maize and Wheat Improvement Center (CIMMYT), this concern is reflected in a special research project on the collection, conservation, evaluation, and equitable sharing of genetic resources for our mandate crops, primarily maize and wheat. As part of that project, we have undertaken to develop methods for identifying the fundamental economic issues related to genetic diversity in crop species, with special reference to developing countries. In consultation with researchers from many disciplines and organizations, we have sought to assemble the current knowledge on these issues for two reasons: to learn from each other's work and to communicate with greater precision about the economics of genetic diversity. This book is one product of that effort. It presents the results of initial economic investigations of diversity in the world's three major food crops: wheat, maize, and rice.

In several ways, this volume furthers our understanding of the economic context in which crop breeders make use of genetic resources and their diversity. First, the authors provide an annotated catalog of the tools used to measure and value genetic diversity, describing their limitations as well as the insights they can offer. This is a useful point of departure for the analyses presented throughout the book and a practical guide for those who are unfamiliar with diversity issues.

Second, the book explores fundamental questions related to the value and efficiency of conserving seed *ex situ*, in gene banks. What are the economics of storing seed in gene banks? How do we assess the economic efficiency of searching for different types of genetic resources to use in breeding programs? Does it make economic sense to store many accessions in a gene bank, even if most accessions are rarely used? The answers to these questions should be extremely useful in managing the *ex situ* conservation of genetic resources.

Questions about genetic diversity are not restricted to gene banks, of course. Increasingly, conservation focuses on the prospects for maintaining diversity *in situ*— in other words, in farmers' fields. A third major contribution of this book is that it examines the many ways that diversity issues are manifested at the farm level in the developing world. Three chapters analyze farmers' objectives and incentives for conserving crop genetic resources in centers of crop diversity, where farmers in some cases already practice *de facto* conservation. These chapters present concrete steps for monitoring, predicting, and developing potential mechanisms to encourage the conservation of crop genetic resources in farmers' fields.

Other contributions of this book include a review of methodological issues that are important for studying the economics of crop species diversity in farmers' fields. For example, how do genetic resources and diversity affect production outcomes for farmers? How do farmers' production choices and constraints affect diversity? What is the value or impact of maintaining diversity in farmers' fields? In addition, several chapters examine how diversity is mediated by policies and institutions. The authors discuss how genetic diversity may be affected by the capacity and organization of the crop research system and by incentives in commercial seed markets. They also consider how new developments in biotechnology might alter research and development processes, with consequences for crop species diversity in developing countries. Finally, they describe the ways that policy may influence farmers' demand for diversity in the crops they grow.

In summary, this book helps to widen, inform, and lend coherence to an emerging area of inquiry that has important implications for CIMMYT and many other institutions concerned about the diversity of crop species. In fact, it reveals how much additional research must be done on these complex issues. We hope that the chapters that follow will stimulate the additional research necessary to understand and protect diversity in the crops that feed the world.

Timothy G. Reeves, Director General, CIMMYT
Prabhu L. Pingali, Director, CIMMYT Economics Program

ACKNOWLEDGMENTS

The authors of this book are grateful for the assistance of several staff of the International Maize and Wheat Improvement Center (CIMMYT). Angélica de la Vega provided administrative and secretarial assistance. Maria Delgadillo helped prepare the manuscript for layout. CIMMYT designers Eliot Sánchez and Marcelo Ortiz produced a camera-ready layout under extreme time pressure and with great precision, thanks to the efforts of Miguel Mellado. Kelly Cassaday managed production.

Much of the research described in this book was first presented at a conference hosted by Stanford University on behalf of CIMMYT, "Building a Basis for the Economic Analysis of Genetic Resources and Diversity in Crop Plants," 17–19 August, 1997, Palo Alto, California. We are indebted to Walter Falcon, Director of the Institute for International Studies at Stanford University, and Chairman of CIMMYT's Board of Trustees, for making that conference possible. We also thank Prabhu Pingali, Director of the CIMMYT Economics Program, whose guidance and encouragement have enabled this research to reach a wider audience through the publication of this book.

I

Introduction

1 FARMERS, GENE BANKS, AND CROP BREEDING: INTRODUCTION AND OVERVIEW

M. Smale, M. R. Bellon, and P. L. Pingali

1.1. BACKGROUND

There is no novelty in the idea that biological diversity among plant and animal species and genetic diversity among subspecies are essential to scientific advance and the future of human society. The fundamental importance of genetic diversity to crop improvement and agricultural systems has long been evident to plant breeders, conservators, agronomists, entomologists, and pathologists. To improve a crop or enhance its resistance to environmental stresses and disease, plant breeders have relied on the genetic diversity in their working collections, on collections of genetic resources stored in gene banks, and on those varieties maintained and selected by farmers. Today, biotechnology and genetic engineering promise to break the biological barriers imposed on conventional plant breeding by incorporating desirable genes and gene complexes from wild relatives and across species.

Economists have only recently turned their research attention to the analysis of crop diversity. Their interest stems from a popular perception that biological scarcity, metered in the declining numbers and complexity of plant and animal species, is an inevitable by-product of technological advance. The acceptance of the International Undertaking on Plant Genetic Resources at the Food and Agriculture Organization of the United Nations in 1983, and the Convention of Biological Diversity at the Rio Earth Summit in 1992, both represented major benchmarks in the recognition by international political fora of the significance of biodiversity.

This volume assembles initial economic investigations of crop diversity in the world's three major food crops (rice, wheat, and maize). Virtually all of the studies summarize research conducted in the field by applied economists with other social scientists and crop scientists, lending the book an interdisciplinary dimension which is essential to the topic. Diversity in rice (*Oryza sativa*), wheat (*Triticum* spp.), and maize (*Zea mays* L.) is explored in the context of the cropping and crop improvement systems of the developing world. These three crops constitute about 85% of global cereals production, and the vast majority of the farmers who produce them live in developing countries. While the experience of industrialized nations and other crops furnishes informative points of comparison, the issues discussed here have historically emerged with the greatest salience in developing countries and with reference to the "green revolution" varieties. Most of the chapters are written by researchers who work in or are closely associated with the international agricultural research centers for rice, wheat, and maize, known as the International Rice Research Institute (IRRI) and the International Maize and Wheat Improvement Center (CIMMYT).

Various perspectives are represented and levels of analysis undertaken in this volume, but measurement techniques, tools, and economic concepts recur from chapter to chapter. Assembling the studies in one volume enables those pursuing further research on this or related topics to assess and build on emerging methods and approaches. Although the book is empirically based, its chapters carry messages for further developments in both the theoretical and applied economics of biological and crop diversity.

1.1.1. Themes

Concerns about crop genetic resources are clustered around two prominent themes. The first is the effect of widespread cultivation of similar genotypes on current crop productivity. Uniformity in the genes conferring resistance to plant diseases can heighten the vulnerability of a crop to mutations in pathogens and to epidemics. When plant disease epidemics occur, they require chemical interventions to avert crop losses. In developing countries where physical infrastructure tends to be poorer and smaller-scale producers are more dispersed, such "curative" interventions can be costly. Notably, although concerns about crop vulnerability have focused on modern varieties, large-scale cultivation of crop populations that are uniform in some genes can and has occurred historically (Nagarajan and Joshi, 1985) among traditional as well as modern varieties (NRC, 1972). The distribution of varieties over geographical areas and time is a causal factor in the mutation of pathogens and the spread of plant disease.

The second theme in the concern for crop diversity is the need to protect potentially valuable alleles and gene complexes that are endangered by "genetic erosion." Economists generally approach conservation issues with a utilitarian, anthropocentric perspective, in which species and subspecies are valued through their use to humans. Use value may include the knowledge that a habitat or species exists. For crop genetic resources, however, compared to an endangered species, the focus of concern is to ensure that valuable alleles and gene complexes are available to enhance crops in the future. A second distinction between questions of genetic erosion in crop species versus endangered species is that the stock of crop genetic resources is not diminished by harvesting the crop. Farmers determine the survival of crop populations and gene complexes when they choose varieties.

The threat of genetic erosion has led to public investments in conservation. The most accepted approach to crop genetic resource conservation is to collect germplasm and store it in *ex situ* gene banks, away from its place of origin, sheltered from natural and human calamities. For wheat, maize, and rice, gene banks are refrigerated vaults designed to catalog and protect seeds for future use. *Ex situ* methods of storage are static, removing crop populations from the evolutionary processes through which they adapt to changes in the human and natural environment. Future advances in crop improvement are predicated on this adaptive potential. Some critics contend that gene banks have excess capacity, while others argue the need to expand the gene bank system.

An alternative paradigm for conservation of crop genetic resources is offered by the *in situ* approach. Generally applied to the preservation in protected areas of wild, pasture, or forest species, *in situ* conservation aims to protect the genetic integrity of plant populations by ensuring their continued evolution within the natural environment (Ford-Lloyd and Jackson, 1986). As compared to preservation of wild species in reserves, on-farm conservation of cultivated plants subjects them to both human and natural selection pressures. It also depends on the active participation of farmers. On-farm conservation is increasingly viewed as serving a complementary function to *ex situ* conservation in centers of crop domestication and diversity (Maxted, Ford-Lloyd, and Jackson, 1997).

The two themes of productivity and conservation merge in concerns over centers of ·crop domestication and diversity, where farmers continue to cultivate and improve their own traditional varieties using historical practices of seed selection and management. Traditional varieties are often understood as evolving, locally adapted, heterogeneous crop populations with special genetic structures, although some researchers question these precepts (Wood and Lenné, 1997). Conservationist concerns pivot on the assumption that even in centers of crop domestication and diversity, modern varieties will inevitably replace traditional crop populations. In fact, the areas of the world where modern maize, rice, and wheat varieties have completely replaced traditional varieties are fairly well defined. In wheat, for example, they include most of Europe, the Mediterranean, China, and the more favored production zones of the Punjabs of Pakistan and India. In many production zones for these crops, farmers practice *de facto* conservation by growing both traditional and modern varieties.

Furthermore, the zones where on-farm conservation programs will have the greatest applicability are likely to be limited in geographical scope. While so-called secondary centers of crop diversity may contain populations with special adaptations and unique genes, the history of human interactions with these populations is relatively brief. If on-farm conservation implies the coherence of a management system for genetic resources rather than the preservation of a specific allele, only some microregions of the world pertain.

Finally, since the cost and difficulty of documenting whether alleles or gene complexes have been "lost" on a large scale would be prohibitive, concerns for "genetic erosion" are still expressed anecdotally. Classifying crop populations is both science and artifice; there is no single system on the basis of which numbers of traditional varieties and landraces can be definitively tabulated.

1.1.2. Previous Research by Economists

The volumes produced by the National Research Council in 1972 and 1993 recorded the concerns of leading scientists about the prospects for and implications of crop genetic uniformity, initiating a wave of interest among researchers in the causes and consequences of the loss of traditional crop populations.

A decade or so after the first NRC volume appeared, books edited by B.G. Norton (1986), E.O. Wilson (1988), and McNeely et al. (1988) included chapters by resource economists about the valuation of biodiversity. The book by Wilson contained the proceedings of the first National Academy of Sciences conference on biodiversity in 1986. These publications were followed by a considerable body of largely theoretical or conceptual pieces by resource economists, often in concert with ecologists and other biological scientists, on the economics of biodiversity and its apparent decline (e.g., Orians et al., 1990; Pearce and Moran, 1994; Swanson, 1995b). A smaller body of economic literature accumulated on the subject of endangered species (Bishop, 1978; Brown and Goldstein, 1984). Several articles by economists on the measurement of biological diversity followed (Solow, Polasky, and Broadus, 1993; Weitzman, 1992, 1993). That research emphasized diversity among species rather than genetic diversity within species.

Economic analysis of genetic diversity issues in agriculture has been slower to appear in the literature, since much of the early research on that subject is found in case studies, only some of which have been published (examples include Brush, Taylor, and Bellon, 1992; Evenson and Gollin, 1997; Heisey et al., 1997; Pray and Knudson, 1994). A compendium that includes an extensive discussion of theoretical issues and empirical methods for assessing the value of crop genetic resources was published in 1998 (Evenson, Gollin, and Santaniello).

The chapters in this book examine several issues that economists have not yet addressed. Chapter 2 classifies and condenses the myriad of diversity indicators available for use by applied economists. The chapters in part II present analyses of the utilization and valuation of crop genetic resources conserved ex situ. Part III consists of three chapters on farmers' incentives for conserving genetic diversity in centers of diversity for rice, wheat, and maize. The studies in part IV investigate the relationship of diversity to crop productivity and stability. In part V, policies affecting the use of genetic diversity in crop breeding programs, the supply of diversity by the research and seed systems, and the demand for diversity by farmers are discussed. A chapter on the relationship of collaborative plant breeding to genetic resource conservation is included as well.

1.2. MAJOR CONCLUSIONS AND UNRESOLVED ISSUES

1.2.1. Do Economists Need New Tools to Study Crop Genetic Diversity?

In general, lending a coherent economic interpretation to popular concerns for genetic erosion and genetic uniformity in the world's three major cereals is not easy. Even though appropriate analytical tools may exist in both the biological sciences and economics considered separately, the interdisciplinary nature of the issues presents methodological challenges. An inherent problem is that because farmers may not perceive or care about the morphological or genetic characteristics used to measure diversity, the relationship between diversity indices and farmers' economic decisions is likely to be indirect and difficult to establish empirically. Genetic phenomena that can be observed only in a molecular biology laboratory are far removed from the daily lives and planting decisions of the world's small-scale farmers, despite claims that biotechnology may serve their interests. Another problem is how to trace, within the confines of the simple statistical methods used in applied economics, the relationship of genes or gene complexes embodied in individual varieties to the volume and variance of crop output produced by large numbers of farmers over geographical expanses. The studies in this volume initiate the development of applied methods that can be used in such work.

1.2.2. Do Economists Have Reliable Indicators to Measure Crop Genetic Diversity?

Diversity in a crop species can be measured by characteristics or traits that are easily observed and enumerated or by those that are not so easy to observe or quantify. The first category includes morphological traits of crop populations sampled in a particular site or the distributions of released varieties over geographical space and time. The second category includes measurements of polymorphism at the molecular level or dissimilarity of pedigrees. While molecular fingerprinting may provide unequivocal evidence of the genetic dissimilarity between crop populations, for economists, both molecular- and pedigree-based indicators share the operational disadvantage that they are unobservable to farmers. In the construction of indices based on named crop populations, an implicit problem is that several taxonomic systems may exist for the same set of populations and merging or reconciling them may not be feasible. The exclusive reliance on named crop populations provides an incomplete picture of crop diversity because it overlooks the diversity within the populations. Depending on the crop and the populations under study, intra-varietal diversity can be low or high relative to inter-varietal diversity.

All measurements, regardless of the special difficulties that each may entail, are based on the capacity of researchers or farmers to group traits, markers, individuals, or populations in order to derive an indicator of distance or dissimilarity. There is no single most accurate indicator of the diversity of crop genetic resources; rather, there are many indicators, of which the most appropriate depends on the objectives and budget of the study under consideration. Each indicator, reflecting the knowledge system on which it is based, views diversity from a single perspective.

1.2.3. Does the Value of Gene Banks Justify Their Cost? What Is Their Optimal Size?

Developing numerical estimates for the value of crop genetic resources conserved *ex situ* remains a guessing game, despite recent theoretical advances. The portion of the use value from crop improvement that is attributable to the gene bank is elusive, given data limitations and the restrictive assumptions that must be employed about the genetic contribution of a variety's ancestors. Furthermore, although discussions of genetic resources commonly invoke such terms from environmental economics as "option value" and "existence value," appropriate definitions and estimation of such values in the empirical context of crop breeding are rare.[1]

The search models developed in chapters 4 and 5 provide new insights but no definitive conclusions regarding the economic value of gene banks and the optimal size of ex situ collections. In chapter 4, perhaps the most counterintuitive finding reported by Gollin, Smale, and Skovmand is that infrequent use of gene bank materials by crop breeders does not imply that marginal accessions have low value. The search for resistance to Russian wheat aphid demonstrates that large collections of landraces may be exhaustively searched only on rare occasions, but these occasions can be associated with big payoffs. The search for resistance to septoria tritici leaf blotch illustrates the point that with conventional search technologies and long research lags, it is efficient for crop breeders to avoid searching among certain categories of genetic resources until they have exhausted their own working collection of elite lines. Chapter 4 also raises questions about the meaning of "utilization"; banks are used not only for searches but also for the advancement of scientific endeavor, and searches often satisfy multiple objectives.

Chapter 5 adds substantively to the literature on the economics of search for crop traits. Evenson and Lemarié develop a two-stage model for the collection and search of genetic resources, analyzing the search for multiple as well as single traits. A model depicting the joint search for multiple traits conforms to the general case of crop improvement in a breeding program; a model of a search for a single trait reflects how crop breeding programs resolve specific, well-defined problems, such as the emergence of new plant diseases and pests. Other search models have underestimated the value of the collection and its optimal size by limiting the analysis to the search for a single (simple or complex) trait in a single period without the collection activity. The authors also show that the optimal size of a collection will be larger if traits are grouped in geographical "niches" rather than distributed randomly, because the best "niche" for each trait must be adequately represented in the collection.

Chapter 3 by Pardey et al. uses the principles of production economics to establish a solid foundation for analyzing the cost of conserving genetic resources in ex situ collections. Pardey et al. provide the first upper-bound estimates of the costs of adding wheat and maize accessions to a gene bank. The estimates are modest enough that the authors conclude that a full-scale estimation of the benefits from conserving additional accessions is not justified. There are substantial differences between the marginal costs of conserving wheat and maize genetic resources, even though they are stored in the same bank in the same location. Fixed costs constitute a high proportion of average total costs, suggesting economies of size in the operation of the gene bank. Scale economies are likely to be introduced through cost complementarities resulting from shared facilities.

None of the studies in part II addresses option value or existence value, although it is not clear what more can be gained by applying these concepts to the assessment of a gene bank's value or the marginal value of a bank accession—unless option value is negative. There is no doubt, however, that further conceptual work on use value as well as research on other crops in other empirical settings is necessary to advance our comprehension of the issues and strengthen policy recommendations. If we concede that storing additional maize and wheat accessions is cheap and estimates of their value are precarious, are further efforts to estimate the marginal value of maize and wheat accessions warranted?

1.2.4. Will Farmers Continue to Grow Traditional Varieties?

De facto conservation is the continued cultivation of traditional varieties or crop populations identified as important genetic resources. The three pieces of research presented in part III are drawn from some of the more detailed, field-based studies that have been conducted expressly to examine the potential for genetic resource conservation on farms. All three chapters ask under what conditions *de facto* conservation of genetic resources occurs.

The opportunity cost of growing traditional varieties changes as agriculture intensifies. Among the communities studied by Bellon *et al.* in northeastern Luzon, Philippines (chapter 6), these changes are mainly associated with trade-offs between the yield and earlier maturity of modern varieties and the consumption qualities of traditional varieties. The authors suggest that breeding interventions to shorten the maturity period of traditional varieties might reduce the opportunity costs of growing them.

Perales, Brush, and Qualset (chapter 7) question two assumptions behind the belief that modern varieties will inevitably replace traditional varieties: modern varieties always yield more, and all farmers maximize profits. In the region of Mexico where these researchers worked, the major traditional varieties of maize grown by farmers compete with each other and with the recommended hybrid either for yield, profits, or both. Maize is not competitive with other crops, however—suggesting that, as is posited in decisionmaking models of the farm household, farmers' objectives encompass more than profit maximization.

Nor can a model of profit maximization explain the fact that farmers grow more than one major variety at the same time. As argued in the microeconomics literature on adoption of new crop varieties, a number of alternative economic hypotheses may explain why farmers grow more than one variety. In their econometric analysis of factors motivating Turkish households to choose among traditional and modern varieties, Meng, Taylor, and Brush (chapter 8) demonstrate that multiple explanations, including missing markets, risk, and agroclimatic constraints, influence the probability that a farm household will grow a traditional variety. An examination of characteristics shared by households with high predicted probabilities of growing landraces further confirms that although risk provides one explanation, other factors, such as district-level average prices and the amount of marketed output, also play a decisive role. In other words, a change in any single economic factor is unlikely to cause farmers to cease growing traditional varieties.

1.2.5. What Is Farmers' Demand for Diversity in the Crop Varieties They Grow?

Bellon *et al.* emphasize that research on farmers' preferences with respect to certain crop populations and the contribution of these populations to genetic diversity in the reference region offers a richer approach to developing on-farm conservation initiatives than the blanket recommendation that farmers should continue to grow traditional varieties. Rice diversity in their study is measured using numbers of named varieties as well as molecular data, for traditional as well as modern varieties of Asian rice, including glutinous and non-glutinous types. Analysis of the molecular data enables Bellon *et al.* to identify a group of rice varieties that is genetically distinct among those grown in the agroecosystems they studied. Combining this information with farmers' perceptions of variety characteristics, Bellon *et al.* propose crop breeding interventions as an incentive for farmers to maintain this special group of varieties.

The chapter by Meng, Taylor, and Brush illustrates some problematic features of modeling farmer demand for diversity through a household model of variety choice. Farmers' choices over crop diversity may be made directly or indirectly, depending upon the crop and whether or not diversity can be observed by the farmer. We might hypothesize that farmers derive no direct utility or economic value from diversity in itself, implying that the level of crop diversity we can measure in the fields of a farm household represents an outcome of other decisions about variety choice, seed selection and management, and crop management. These decisions are in turn motivated by economic factors, as well as exogenous agroecological and household parameters. Such a behavioral formulation dictates a recursive model and estimation procedure.

Like the authors of chapter 8, we might hypothesize instead that unobservable factors, not captured in the systematic portion of the estimating equations, link the conscious decisions of farmers to the diversity outcome. In their econometric analysis, however, Meng, Taylor, and Brush fail to reject the hypothesis that the level of morphological diversity within wheat landraces grown by farm households is independent of their choice of variety. Alternatively, farmers' utility may be defined directly over some dimension of observable diversity in the crop, such as color or grain texture. Yet modeling and testing empirically the endogeneity of these variables can be awkward because units of observation do not conform, definitions conflict, or data requirements are prohibitive. Nor is modeling the diversity outcome in and of itself a straightforward exercise. Many of the theoretical parameters related to human and natural selection pressures that are known to shape the evolution of genetic structure in crops cannot be picked up in a cross-sectional survey or a small panel data set of the kind most often analyzed by applied economists.

1.2.6. Can Collaborative Plant Breeding Provide an Incentive for On-Farm Conservation?

Chapter 15 explores some of the economic issues related to the notion that collaborative plant breeding can provide an incentive for conserving genetic resources on the farm. Collaborative (or participatory) plant breeding uses the skills and experience of both farmer-breeders and professional plant breeders to improve crop plants. The extent of participation by farmer- and professional breeders depends on the case. At one extreme, farmers may be brought onto the experiment station to identify varieties or populations from among those being developed by breeders. At the other extreme, farmers may be directly involved in the elaboration of breeding strategies or modified seed selection practices, which enable them to improve their own crop populations *in situ*. Some of these farmers may then serve as local seed suppliers in non-commercial seed systems, backed by a local, national, or regional gene bank.

Proponents of collaborative plant breeding argue that although professional plant breeders have conventionally sought to develop fewer varieties adapted to a wider range of geographical locations, this approach can support the maintenance of more diverse, locally adapted plant populations. The biological validity of this proposition will need to be tested. The question of farmers' incentives to engage in such efforts is crucial, since they are likely to be time-consuming and payoffs may not occur for years.

For crops such as beans and potatoes, social and biological scientists in the international agricultural research system and other institutions have generated a large body of research on collaborative plant breeding, community seed systems, and crop diversity.[2] The case studies for maize in Mexico, described in chapter 15, raise doubts that farmers will be able to reap the benefits from one particular type of breeding intervention—modified mass selection practices—unless certain technical issues are better articulated, farmers' seed management practices are better understood, and farmers' own perceptions and knowledge systems are properly taken into account. In each case, biological and social issues are closely intertwined, and failure to consider one or the other leads to an incomplete assessment of the problem.

1.2.7. What Is the Effect of Diversity on Crop Productivity and Stability?
When the crop output has a market value and when a supply response to changes in the quantities of inputs can be observed, a hedonic approach and statistical methods can be used to estimate the value of the contribution of inputs that are traded and those that are not, as well as certain attributes of the output.[3] In a sense, any study investigating the impact of plant breeding on yield is analyzing the effects of genetic resources on productivity, broadly defined. One of the first studies that addressed the valuation of particular attributes of genetic resources, such as the complexity of their pedigrees, was conducted by Gollin and Evenson in 1990 and published in 1998.

Both Widawsky and Rozelle (chapter 10) and Hartell et al. (chapter 9) use conventional production function analyses to investigate relationships among crop diversity, productivity, and stability. Widawsky and Rozelle employ a generalized Cobb–Douglas production function with a stochastic specification (Just and Pope, 1979). The authors measure rice diversity using a Solow/Polasky distance index constructed from pedigree data (see chapter 2) and data from townships in Zhejiang and Jiangsu Provinces, eastern China. Although regression results are mixed, they suggest that at the township level, varietal diversity may be negatively associated with both the mean and variance of rice yields. As implied by the mean–variance model, policymakers who choose to emphasize crop stability must pay a price in terms of productivity.

Using district-level data for wheat production in the Punjab of Pakistan, Hartell et al. tested the effects of a set of diversity indicators by estimating a Cobb–Douglas production function for the mean production relationship and a separate yield stability equation. The indicators represent different genetic resource and diversity concepts, including two variables representing the crossing history and landrace ancestry of the variety, as well as three variables representing spatial, temporal, and latent diversity. Latent diversity was measured using pedigrees and calculated with the Weitzman distance index (see chapter 2). Hartell et al. conclude that genealogical distance may enhance yield stability at the district level. In the irrigated areas, lower spatial diversity is associated positively with mean yields, while lower temporal diversity has a depressing effect. In other words, more recent releases have higher yields, and when more farmers grow them over greater areas, district-level yields are higher. In the rainfed areas, the genealogical distance among varieties and the number of different landraces in their genetic background are positively associated with yield. This finding probably reflects the breeding history of the varieties that perform better in this environment.

Smale and Singh (chapter 11) relate genetic profiles, or data on the presence of specific genes and gene complexes in individual varieties, to yield losses from leaf rusts in modern bread wheat varieties. In contrast to the use of genealogical information with data on aggregate crop production, the use of genetic profiles with trial data provides results that are relatively easy to interpret. This type of data generates only limited information about the magnitude of actual yield losses from disease in farmers' fields, however. To complete the analysis of economic impact, the authors used simulation. They conclude that diversifying genetic resistance to leaf rusts in modern wheats by accumulating non-specific genes was worth the investment.

Several methodological issues are raised by the chapters in part IV. The first is that an empirical relationship between pedigrees and crop performance is difficult to interpret meaningfully, because even the theoretical relationship is indirect. This is especially true outside the experiment station, when data are aggregated to township, district, or regional levels. Furthermore, capturing the effects of particular genes or gene complexes with the crop production data and estimation methods typically used by applied economists may not be feasible because of the overwhelming effects of crop management practices and environmental conditions on yield.

Finally, the concepts of spatial and temporal diversity, and latent diversity based on genealogies, are to some extent another way of summarizing differences among modern varieties in terms of their comparative breeding history, popularity among farmers, and release date. At an aggregated level of analysis, how can the effects of economic as well as genetic variables be estimated jointly? The task for future work on the relationship of crop diversity to productivity and stability will be to develop a fuller specification of an economic decisionmaking model with genetic indicators in such a way that policy implications are more transparent and informative.

1.2.8. Is the Development and Use of Biotechnology in Developed Countries Instructive for Developing Countries?

Biotechnology provides tools to assure the continuous advances in crop productivity that are necessary to counteract growing human populations and resource degradation in developing countries. The experience of the past 30 years in the United States shows only modest achievements, however. As indicated in chapter 12, biotechnology is a promise rather than a panacea, and a nexus of institutional arrangements as well as related crop technologies is necessary to support its effective utilization. Its emergence has generally occurred where a critical mass of expertise in natural, technological, social, and management sciences is complemented by substantial financial and entrepreneurial resources; its utilization is generally accompanied by privatization, closer links between universities and industry, product differentiation and contracting, and expansion in the range of agricultural products.

Large biotechnology companies in developed countries may have little interest in solving problems specific to developing countries. Biotechnology, like any technology, requires local adaptation. As a consequence, even when developing countries do not count themselves among the leaders of biotechnology research, they may want to invest in related human and industrial capital. Similarly, global investment in public research on the genetic enhancement of varieties oriented toward the needs of subsistence farmers is essential, and it must be buttressed by free or low-cost access by crop breeders to intellectual property rights and other technology components. On the other hand, agreements that provide preferential treatment of genetic materials "for the poor" should be examined to prevent their abuse. Biotechnology material used in developing countries to produce export goods should not be exempt from the payment of intellectual property rights. Finally, developing countries should make efforts not only to protect their "biodiversity" but to more fully understand the meaning and implications of the term. Royalties do not provide long-term economic gains unless wisely invested.

1.2.9. How Is Genetic Diversity Affected by International Coordination of Research Roles?

The international agricultural research systems for rice and wheat enable crop breeding programs in developing countries to incorporate sources of genetic diversity from beyond national boundaries into the cultivars grown locally. Traxler and Pingali (chapter 13) outline the roles of the institutions in these two systems and how they have shifted since the late 1960s (the early years of the green revolution). The sheer number of materials released globally, through the efforts of national as well as international agricultural research institutions, has augmented continually for both rice and wheat since the early phases of the green revolution.

The research contributions of collaborating institutions have evolved in distinct ways for the two crops. As national programs have come to rely less on IRRI for strains of rice that are nearly ready for release, IRRI has gradually diverted resources into upstream research and germplasm exchange. ("Upstream research" can be thought of as strategic research and research that taps novel sources of genetic diversity—the precursors to future productivity gains.) By contrast, national collaborators in the wheat research system still depend heavily on CIMMYT to produce nearly finished varieties, curbing opportunities for CIMMYT to shift research resources toward more strategic research. The strain on resources is especially acute in an era of shrinking research budgets, with damaging implications for the future of the international wheat research system.

1.2.10. How Do Incentives in Commercial Seed Markets Affect Crop Diversity?

In chapter 14, Morris and Heisey carry the institutional analysis one step further in the crop improvement process, into the seed systems that support the diffusion of new varieties among the farmers who demand them. The analysis begins with an overview of data on temporal, spatial, and latent diversity among modern varieties of wheat, rice, and maize. Wide variation in temporal diversity is apparent among regions but is generally greater in wheat and rice than in maize. The "age" of varieties is defined over a different metric in maize than in the self-pollinated crops, since the genetic structure of a maize variety changes more rapidly under field conditions. The authors find no clear relationship between latent diversity in modern varieties, as measured by genealogies, and either the crop species or the economic development of the country. They conclude that while the concentration of area among modern varieties is cyclical, most studies show that spatial diversity has improved rather than worsened over time.

Morris and Heisey then summarize the incentives for varietal diversification among the farmers who demand seed and in the industry that supplies it, arguing that utility-maximizing behavior on the part of farmers and seed producers is unlikely to result in a level of crop diversity in farmers' fields that reflects the social optimum.[4] Several policy options are available to push the private optimum closer to the social optimum, although the authors suggest that the alternatives may be more workable on the supply side than on the demand side. The issues raised by Zilberman, Yarkin, and Heiman (chapter 12) and Traxler and Pingali (chapter 13) may be particularly relevant, since those issues operate through the supply of genetic resources.

1.2.11. Do We Know Enough to Design On-Farm Conservation Programs?

Any research dealing with the feasibility, costs, and benefits of on-farm conservation must address a number of key issues raised in chapters 6, 7, 8, 14, and 15: Which farmers should conserve? Which crop populations should be conserved? Which policy instruments or technical interventions provide the most effective incentive mechanisms for conservation? Which conservation goal should be pursued? Which indicators of genetic diversity should be used?

The policy issue addressed in the work by Meng, Taylor, and Brush concerns the identification of farm households that are most likely to maintain diversity; these are the households it would cost the least to target in a conservation program. Bellon *et al.* identify which crop populations contribute most to genetic diversity in a reference population and are of greatest utility to farmers; these are suitable candidates for on-farm conservation because they are desirable genetic resources, and farmers are more likely to maintain them. Perales, Brush, and Qualset distinguish major from minor crop populations, based on the scale of cultivation. The most suitable means of encouraging the cultivation of major crop populations might be improvements in farmers' breeding procedures, without particular concern for the introduction of new genes or loss of alleles. For minor crop populations, locally based seed-saver schemes could be more effective.

The detailed studies of *de facto* genetic resource conservation by farmers in Mexico, Turkey, and the Philippines provide not only rich empirical analyses but conceptual frameworks and an economic model for analyzing the relationship between farmers' choice of variety and crop diversity. This understanding of farmer decisionmaking and crop diversity at the microeconomic level is not matched, however, by a comprehension of the institutional mechanisms that are required to implement on-farm conservation programs. Clearly, further work is needed to address more specifically the nature of the institutional arrangements, incentives, and agents that would need to be involved in initiatives for on-farm conservation.

Temporal dimensions, such as crop evolution and changes in the parameters of utility functions, cannot be captured in detailed cross-sectional studies. Other detailed, cross-sectional case studies of this type will need to be followed by longitudinal studies in order to strengthen policy recommendations. Further, there is extensive theoretical debate about which genetic diversity indices (see chapter 2) are good indicators of the evolutionary potential of a crop population.

As mentioned by Morris and Heisey in chapter 14, establishing a relationship between the varieties chosen by individual farmers and varietal distributions across space and time, or between levels of economic analysis, is crucial. Other levels might include not only analyses of the seed industry as a whole, but also tests of hypotheses about the roles of culture and community in the decisions of individual farmers.

Finally, goals need to be articulated more fully. Like Morris and Heisey (chapter 14), Smale *et al.* (chapter 15) frame their discussion in terms of the likely divergence between the utility-maximizing behavior of farmers and the social optimum for genetic resource conservation on farms. Soleri and Smith (1995) have described this dilemma as the contrast between farmer and conservationist perspectives. Each perspective implies a distinct goal in terms of genetic diversity. The goal from the conservationist perspective is to maximize the allelic diversity in targeted crop populations; the goal from the farmer perspective is to maintain the genetic diversity that enables farmers to best meet their own current and future production needs as they perceive them. This adds another wrinkle to the "social trap" problem raised in chapter 14.

1.3. FINAL COMMENTS

As this overview has shown, the economic questions raised by concerns about crop genetic resources are as fascinating as they are important to the future of global society. The approaches required to answer them, although they follow well-traveled routes of inquiry, present special challenges because of their interdisciplinary nature. This book represents a first step in elaborating concepts and tools for applied economic investigations in the diversity of crop genetic resources. As well as making a concrete advance toward elucidating this area of research, the chapters that follow call for further development of both methods and theory to examine the unresolved issues in greater depth.

Notes
1 Brown (1990), Swanson (1995a), and more recently Evenson, Gollin, and Santaniello (1998) have summarized theoretical and empirical approaches to valuing crop genetic resources. Evenson and Gollin (1997) were perhaps the first to estimate the value of marginal accessions, based on a study of IRRI's gene bank. Fisher and Hanemann (1984) used the case of teosinte in Mexico primarily as a means of proving the existence of positive option value, in the sense defined by Arrow and Fisher (1974) and Henry (1974). A recent paper by Cooper (1998) presents a model for measuring the economic returns to publicly funded investments in genetic resource conservation *ex situ* and *in situ*. The approach considers the effects of uncertainty as well as the value of avoiding irreversible decisions, but the data needed for its estimation are extensive.
2 For example, see Eyzaquirre and Iwanaga (1996); CGIAR (1997); Sperling, Scheidegger, and Buruchara (1996); and Sperling and Loevinsohn (1996).

3 In several chapters in Evenson, Gollin, and Santaniello (1998), hedonic valuation techniques are used to value genetic resources as a producer good. While the editors argue that this approach offers perhaps the most convincing measure of the value of genetic resources by directly linking their attributes to crop output, they also note that a complete set of data on total factor productivity in India was required to disentangle the value of the genetic material from the value of other research inputs and changes in factor use (Gollin and Evenson, 1998).

4 Termed the "social trap" (Schmid, 1987), this problem has been treated formally in models of impure public goods (Cornes and Sandler, 1986) and is applied to wheat diversity in Heisey et al. (1997).

References

Arrow, K., and Fisher, A. 1974. Environmental preservation, uncertainty, and irreversibility. *Quarterly Journal of Economics* 88: 312–319.

Bishop, R. C. 1978. Endangered species and uncertainty: The economics of a safe minimum standard. *American Journal of Agricultural Economics* 60: 10–18.

Brown, G. M. Jr. 1990. Valuation of genetic resources. In G. H. Orians, G. M. Brown, Jr., W. E. Kunin, and J. E. Swierzbinski (eds.), *The Preservation and Valuation of Biological Resources*. Seattle: University of Washington Press.

Brown, G. M. Jr., and J. H. Goldstein. 1984. A model for valuing endangered species. *Journal of Environmental Economics and Management* 111: 303–309.

Brush, S., J. E. Taylor, and M. Bellon. 1992. Technology adoption and biological diversity in Andean potato agriculture. *Journal of Development Economics* 39: 365–387.

CGIAR (Consultative Group on International Agricultural Research) System-Wide Project (ed.). 1997. *New Frontiers in Participatory Research and Gender Analysis: Proceedings of the International Seminar on Participatory Research and Gender Analysis for Technology Development*. Cali: CGIAR System-Wide Project.

Cooper, J. 1998. The economics of public investment in agro-biodiversity conservation. In R. E. Evenson, D. Gollin, and V. Santaniello (eds.), *Agricultural Values of Plant Genetic Resources*. Wallingford: CAB International.

Cornes, R., and T. Sandler. 1986. *The Theory of Externalities, Public Goods, and Club Goods*. Cambridge: Cambridge University Press.

Evenson, R. E., and D. Gollin. 1997. Genetic resources, international organizations, and improvement in rice varieties. *Economic Development and Cultural Change* 45: 471–500.

Evenson, R. E., D. Gollin, and V. Santaniello (eds.). 1998. *Agricultural Values of Plant Genetic Resources*. Wallingford: CAB International.

Eyzaguirre, P., and M. Iwanaga (eds.). 1996. *Participatory Plant Breeding. Proceedings of a Workshop on Participatory Plant Breeding 26–29 July 1995, Wageningen, the Netherlands*. Rome: International Plant Genetic Resources Institute (IPGRI).

Fisher, A. C., and W. M. Hanemann. 1984. *Option Value and the Extinction of Species*. Giannini Foundation Working Paper 269. Berkeley: Department of Agricultural and Resource Economics, University of California at Berkeley.

Ford-Lloyd, B. V., and M. T. Jackson. 1986. *Plant Genetic Resources: An Introduction to Their Conservation and Use*. London: Edward Arnold.

Gollin, D., and R.E. Evenson. 1998. An application of hedonic pricing methods to value rice genetic resources in India. In R. E. Evenson, D. Gollin, and V. Santaniello (eds.), *Agricultural Values of Plant Genetic Resources*. Wallingford: CAB International.

Heisey, P. W., M. Smale, D. Byerlee, and E. Souza. 1997. Wheat rusts and the costs of genetic diversity in the Punjab of Pakistan. *American Journal of Agricultural Economics* 79: 726–737.

Henry, C. 1974. Investment decisions under uncertainty: The irreversibility effect. *American Economic Review* 64: 1006–1012.

Just, R. E., and R. D. Pope, R. 1979. Production function estimation and related risk considerations. *American Journal of Agricultural Economics* 61: 276–284.

Maxted, N., B. V. Ford-Lloyd, and J. G. Hawkes (eds.). 1997. *Plant Genetic Conservation: The In Situ Approach*. London: Chapman and Hall.

McNeely, J. A. (ed.). 1988. *Economics and Biological Diversity: Developing and Using Economic Incentives to Conserve Biological Resources*. Cambridge: IUCN Publication Services Unit.

Nagarajan, S., and L. M. Joshi. 1985. Epidemiology in the Indian subcontinent. In A. P. Roelfs and W. R. Bushnell (eds.), *Diseases, Distribution, Epidemiology, and Control*. Vol. 2 of *The Cereal Rusts*. London: Academic Press.

NRC (National Research Council). 1972. *Genetic Vulnerability of Major Crops*. Washington: National Academy of Sciences.

Norton, B.G. (ed.). 1986. *The Preservation of Species: The Valuation of Biological Diversity*. Princeton: Princeton University Press.

Orians, G. H., G. M. Brown, Jr., W. E. Kunin, and J. E. Swierzbinski (ed). 1990. *The Conservation and Valuation of Biological Resources*. Seattle: University of Washington Press.

Pearce, D. and D. Moran. 1994. *The Economic Value of Biodiversity*. London: Earthscan.

Pray, C. and M. Knudson. 1994. Impacts of intellectual property rights on genetic diversity: The case of U.S. wheat. *Contemporary Economic Policy* 12: 102–111.

Schmid, A. A. 1987. *Property, Power, and Public Choice*. Second edition. New York: Praeger.

Soleri, D., and S. E. Smith. 1995. Morphological and phenological comparisons of two Hopi maize varieties conserved *in situ* and *ex situ*. *Economic Botany* 49:56–77.

Solow, A., S. Polasky, and J. Broadus. 1993. On the measurement of biological diversity. *Journal of Environmental Economics and Management* 24: 60–68.

Sperling, L., and M. Loevinsohn (eds.). 1996. *Using Diversity: Enhancing and Maintaining Genetic Resources On-Farm, Proceedings of a workshop held on 19–21 June, 1995, New Delhi*. New Delhi: International Development Research Centre (IDRC).

Sperling, L., U. Scheidegger, R. Buruchara. 1996. *Designing Seed Systems with Small Farmers: Principles Derived from Bean Research in the Great Lakes Region of Africa*. Agricultural Research and Extension Network Paper No. 60. London: Overseas Development Agency.

Swanson, T. 1995a. The values of global biodiversity: The case of PGRFA. Paper presented at the Technical Consultation on Economic and Policy Research for Genetic Resource Conservation and Use, June 21–22, International Food Policy Research Institute, Washington, DC. Mimeo. Cambridge, UK: Cambridge University, Faculty of Economics.

Swanson, T. (ed.) 1995b. *The Economics and Ecology of Biodiversity Decline*. Cambridge, UK: Cambridge University Press.

Weitzman, M. L. 1992. On diversity. *Quarterly Journal of Economics*. 107: 363–406.

Weitzman, M. L. 1993. What to preserve? An application of diversity theory to crane conservation. *Quarterly Journal of Economics* 108: 1557–1583.

Wilson, E.O. (ed.) 1988. *Biodiversity*. Washington, DC: National Academy Press.

Wood, D. and J. Lenné. 1997. The conservation of agro-biodiversity on-farm: Questioning the emerging paradigm. *Biodiversity and Conservation* 6: 109–129.

2 DEFINITION AND MEASUREMENT OF CROP DIVERSITY FOR ECONOMIC ANALYSIS

E. C. H. Meng, M. Smale, M. Bellon, and D. Grimanelli

2.1. INTRODUCTION

The study of biodiversity, or biological diversity, generally refers to a broad area of scientific inquiry that encompasses all living organisms and their relationship to each other. The studies described in this volume analyze the diversity of crop genetic resources within the context of economic behavior and decisionmaking, focusing specifically on the plant populations of three major food crops (rice, wheat, and maize) among domesticated species. Sections of the book explore the conservation of crop genetic resources and their diversity both *ex situ* and *in situ*, the impacts of diversity on the yield and yield stability of a crop outside the experiment station, and the effects of policies on the utilization of these resources by farmers and scientists.

All of the studies in this book investigate diversity within a single crop species. Here, we use the term "crop diversity" to refer to any one of several different but related concepts. In most of the chapters, "crop diversity" describes the observable variation in a particular plant feature or set of features either within or among distinct crop populations of the same species, as these populations are identified in fields by farmers or scientists. In other chapters, the term designates the variation that can only be identified using molecular techniques in the laboratory or knowledge of the crossing history and ancestors of the crop populations under study.

Successful economic analysis of any of the issues related to the diversity of crop genetic resources requires that the concept of diversity be well defined and that the measurement technique be appropriate to the type of analysis and its objectives, given data, budget, and/or time constraints. In the decision to focus economic research on a particular issue, the use of more than one concept of genetic diversity may be feasible or even appropriate. Each concept will describe or classify diversity differently, and none can be deemed correct or incorrect *a priori*. The appropriateness of the concept that is chosen is largely a function of the objectives of the study and of the level at which the analysis takes place. For example, the concept of diversity developed for a study analyzing the determinants of on-farm crop diversity at the farm household level is distinct from that applied in a regional study of the relationship between the demand for and supply of modern varieties and crop diversity. Both of these will differ from the concept of diversity employed to study genetic resource utilization in a gene bank.

It is also important to distinguish between the concept of diversity and the measurement tool. The measurement tool is simply a mathematical construct that enables a given concept of diversity to be incorporated more conveniently in an analytical model. Once a concept has been chosen, an index allowing it to be expressed as a number or scalar that varies over the units of observation can be calculated. Many of these indices are drawn from ecological and agronomic literature and are developed from some kind of distance metric. Some of the most commonly used means of conceptualizing and categorizing diversity, as well as indices associated with their use, are discussed in this chapter and summarized in Table 2.1.

2.2. DEFINING THE CROP POPULATION UNDER STUDY

Questions regarding the appropriate definition of the crop population emerge in the construction of almost any empirical index, and their resolution depends on the objectives of the study as well as the nature of the crop. For example, crop populations may be defined as they are named by local farmers. These are sometimes called "farmer varieties" or "folk varieties." In other cases, they may be classified by professional germplasm collectors or crop scientists employed by private companies or national or international agricultural research institutions. Crop scientists use terms such as "race," "population," "cross," and "line," as well as "variety," where "variety" typically refers to the finished product of a crop improvement program. Local agricultural officials may classify crop populations based on some other criterion for the purposes of an agricultural survey or census, and their classification system and its meaning may change between reporting periods when policymakers demand a new type of information. Although a set of farmers, scientists, or officials may distinguish crop populations using classification systems, or taxonomies, that are internally consistent, any pair of taxonomies can be difficult to compare, merge, or unify.

Typically, economists studying agricultural research systems and adoption of new crop varieties have relied on a simplistic distinction between modern and traditional varieties, in which "modern variety" refers to a specific product (as in the case of semidwarf wheats) or a general product (as in the case of maize hybrids, synthetics, and composites) of formal plant breeding programs, and "traditional variety" refers to a product of natural and farmer selection. Both products are mutable over their lifetimes in farmers' fields because of human and environmental pressures. The likelihood of mutation is a social and biological fact with implications for both genetic diversity and economics. Genetic changes, sometimes negatively termed "genetic deterioration" or "contamination," are more rapid with open-pollinated crops (e.g., maize) than with self-pollinated crops (e.g., rice and wheat), but they occur in both crop types. Modern varieties that farmers and nature have selected over a period of years are sometimes called "rusticated," "creolized," or, more positively, "locally adapted." Several of the chapters in this book develop more subtle classifications than "modern" and "traditional," although such classifications can easily become unwieldy. The problem of defining the populations in order to count, classify, discriminate among them, or cluster them together extends to most of the concepts and empirical indices described in this chapter.

Table 2.1. Crop Diversity Concepts and Indices Used in This Volume

Chapter	Crop	Population	Level of analysis	Concept	Measurement	Index
4	Wheat	Accession	Gene bank	Apparent	Probability distribution of trait across accessions	–
6	Rice	Modern and traditional varieties	Household, ecosystem	Inter-varietal, intra-varietal, latent, apparent	Molecular, traits reported by farmers	Nei
7	Maize	Modern and traditional varieties, races	Household, community	Inter-varietal, spatial, apparent	Counts, areas	–
8	Wheat	Modern and traditional varieties	Household	Intra-varietal, inter-varietal, apparent	Morphological traits measured by scientists	Shannon–Wiener
9	Wheat	Modern varieties	District	Inter-varietal, temporal, spatial, latent	Age, area, pedigrees	Herfindahl, Weitzman
10	Rice	Modern varieties	Township	Inter-varietal, spatial, latent	Area, pedigrees	Area-weighted, Solow/Polasky
11	Wheat	Modern varieties	Breeding program	Intra-varietal, latent	Genes	–
14	Rice, wheat, maize	Modern varieties	Regional, national	Inter-varietal, spatial, temporal, latent	Area, age, pedigrees	–
15	Maize	Traditional varieties, seed lots	Household, community	Inter-varietal, intra-varietal, temporal, spatial, latent, apparent	Age, area, molecular, morphology	–

2.3. SPATIAL AND TEMPORAL DIVERSITY

2.3.1. Spatial Diversity

"Spatial diversity" is probably the most commonly recognized concept of diversity and refers to the amount of diversity found in a given geographical area. Spatial diversity implies both richness and abundance (that is, a considerable number of the species or populations under study as well as their extensive distribution) (Magurran, 1991). Measures of richness, such as the number of crop populations reported or collected in an area, are typically the simplest way to assess spatial diversity. Species richness measures, however, are highly dependent on the sample size (Begossi, 1996). Furthermore, a simple count is likely to present a skewed picture of diversity if the crop populations are distributed unevenly. Measures of species abundance or evenness take into consideration the likelihood that not all populations are equally represented. These measures are often estimated mathematically by fitting a statistical distribution (Magurran, 1991).

The selection and use of spatial diversity measures requires caution since the number of crop populations is not necessarily an accurate reflection of genetic diversity. The crop populations that are distinguishable based on one taxonomy may not be distinguishable when another taxonomy is used; two distinct "varieties" may actually be very similar morphologically or genetically. Reliance on named crop populations may either underestimate diversity if those identified by the same name possess important underlying genetic differences, or it may overestimate diversity if populations identified by different names are actually very similar.

2.3.2. Temporal Diversity

"Temporal diversity" refers to the rate of change or turnover of varieties. Duvick (1984) has described it as "genetic diversity in time." The replacement of varieties reduces the potential exposure to disease epidemics resulting from pathogen mutations that overcome the genetic resistance in older varieties. Varietal turnover is important for modern agriculture and in some ways substitutes for the spatial diversity that is characteristic of traditional varieties (Apple, 1977; Plucknett and Smith, 1986). The economically optimal rate of varietal turnover in a given area is jointly determined by a number of factors, including the rate of mutation of disease organisms, the structure of genetic resistance to disease in a variety, and the production environment (Heisey and Brennan, 1991).

2.4. APPARENT DIVERSITY AND LATENT DIVERSITY

2.4.1. Apparent Diversity

"Apparent diversity" refers to the variation in physical characteristics within or among crop populations that is observable in the field by farmers or scientists. Although a count of named varieties has been one of the most common means of denoting apparent diversity in crop genetic resources, the expression of apparent diversity is not limited to the use of named varieties alone. Frequencies based on the specific characteristics and performance of plant populations can be used in a similar manner and decrease the likelihood of overlooking some of the differences that may not be picked up when relying on names. The development of these indices requires that the scope of the study be limited to a well-defined geographical area where it is possible to draw physical samples from crop populations.

Morphological traits are physically observable descriptors often used in the crop science literature to describe plant populations and are thus a natural choice for assessing diversity in crop populations (Louette, Charrier, and Berthaud, 1997; Meng, 1997). These traits can be measured both quantitatively (e.g., height, spike [wheat] or ear [maize] length, thousand-kernel or cob weight) and qualitatively (e.g., kernel or grain color, presence of awns). Since information is usually collected for multiple characteristics, the data are often compiled for ease of use into an index that is partially determined by the type of data collected. However, because observable physical differences could result from either genetic differences or differences in the environment, precautions must be taken to account for interactions between genotype and environment before drawing any conclusions regarding diversity levels. Furthermore, in certain crop populations, the presence of morphological differences may mask the closeness of the actual genetic relationship (Dudley, 1994).

2.4.2. Latent Diversity: Molecular Methods

The presence of genetic diversity among or within crop populations is not necessarily observable in the physical characteristics of the plants growing in the field. The use of molecular markers is increasingly common for the detection of differences at the genetic level among and within crop populations. Scientific interest in any particular crop population or set of populations as candidates for conservation lies principally in their evolutionary potential. There is no test for evolutionary potential, although the absence of genetic diversity strongly implies its absence. One indicator of evolutionary potential is genetic polymorphism. Molecular methods for measuring genetic diversity enable the assessment of polymorphism within and among crop populations.

"Polymorphism" is defined as the simultaneous and regular occurrence in the same population of two or more discontinuous variants or genotypes in frequencies which cannot be accounted for by recurrent mutation (Ford, 1940). Selection pressures, both environmental as well as human, act upon these populations by eliminating less adapted genotypes and maintaining the more adapted ones. Selection affects the allele frequencies across varieties.[1]

Currently, a number of procedures are used to estimate genetic diversity or polymorphism at the biochemical and molecular levels. Electrophoresis, the separation and visualization of allozymes, has been the most frequently used in assessing plant genetic diversity at these levels since it is widely applicable, cost-effective, and relatively rapid (Hamrick and Godt, 1997). Hamrick and Godt (1997) provide an example of its use to assess genetic variation for multiple crops; Doebley, Goodman, and Stuber (1985) do the same for races of maize in Mexico. Advances in biotechnology have made possible new and more sophisticated molecular methods for detecting genetic variation at the DNA level, such as restriction or amplified fragment length polymorphism (RFLP or AFLP) markers and random amplified polymorphic DNAs. Each of these molecular methods involves an examination of the relationship between crop populations or individual materials from band patterns that reveal variation in DNA sequences (Dudley, 1994).

The use of biochemical or molecular markers can determine whether two populations are similar or not by offering a genetic fingerprint of each one. It is important to recognize, however, that the existence of polymorphism at the DNA or isozyme level does not necessarily relate to adaptive or evolutionary potential. Nor does the absence of polymorphism at the molecular level imply a lack of adaptive or evolutionary potential. The relationship between the degree of polymorphism and evolutionary potential also depends on the species. When examined at the molecular level, two distinct lines of wheat might reveal 10% polymorphism, while two related populations of maize may display as much as 90% polymorphism. Scientists would not deduce from these observations, however, that maize has greater evolutionary potential than wheat.

Since these methods all require laboratory time and materials, they can be considerably more costly than the use of methods based only on physical observation. The development of diversity indices from molecular measurements, like those constructed from morphological measurements, requires physical samples. For these indices to be linked to the behavioral decisions of farmers, samples must also be drawn systematically from the crop populations in a well-defined geographical area. Because farmers choose to grow crop populations based on characteristics they can see rather than on the use of genes they cannot see, the linkage between farmers' economic decisions and genetic diversity measured at the molecular level is difficult to establish empirically. That many important economic traits are determined by multiple genes further complicates the linkage.

2.4.3. Latent Diversity: Pedigree Analysis
Genealogical characteristics can also be used to assess the extent of unobservable genetic variability in crop populations. Since genealogies are recorded only for varieties released by crop breeding programs, they can be applied only to modern varieties (a segment of the pedigree for the modern wheat variety Sonalika is presented in Figure 2.1). The absence of known pedigrees for landraces precludes their inclusion in any pedigree-based analysis. Nor can farmers observe the genealogies for any of the modern varieties they grow. Similar to the situation described above for molecular analysis, the relationship between farmers' economic decisions and crop diversity measured by genealogical indices is difficult to establish.

Because pedigrees of varieties are often quite complicated, many different methods have been employed to represent the information contained in the pedigree. Some of them focus on "pedigree complexity": that is, specific details from the pedigrees, such as the numbers and origin of landraces in ancestry or numbers of breeding generations since the first cross (Gollin and Evenson, 1990). Numbers of distinct parental combinations and numbers of unique landrace ancestors per pedigree were used in Smale and McBride (1996) and Hartell (1996), although a theoretical relationship between these numbers and genetic diversity has not yet been established. They are better used as indicators of the relative contributions of certain types of genetic resources to varieties, or the extent of breeding program activity versus farmer selection that generated the variety in question. Similar types of pedigree analysis have been applied by Brennan and Fox (1995) and Pardey *et al.* (1996) to estimate the proportional contribution of germplasm by source to economic returns from research.

Another method for incorporating pedigree information into a usable form is through the calculation of coefficients of diversity (1 – the coefficient of parentage). The coefficient of parentage (COP) is a pair-wise comparison and estimates the probability that a random allele taken from a random locus in a variety *X* is identical, by descent, to a random allele taken from the same locus in variety *Y* (Malecot, 1948). Values range from 0 to 1, with higher values indicating greater relatedness (for historical development of this concept, see Wright, 1922; Malecot, 1948; Kempthorne, 1969; and Cox *et al.*, 1985).

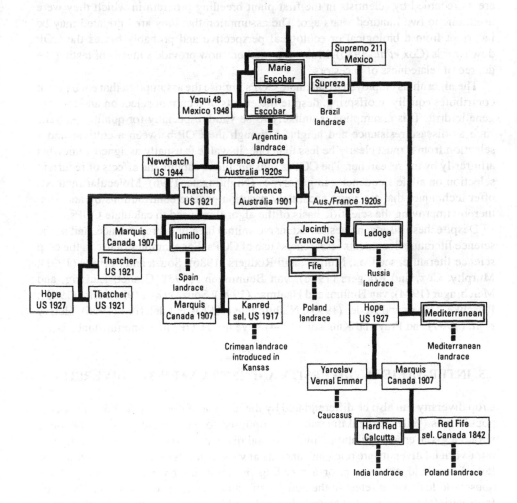

Figure 2.1. Segment of a Bread Wheat Pedigree (cv. Sonalika).

The use of COP analysis reflects "the relative change in diversity due to breeding" among breeding materials over time and contributes to the evaluation of the effect of breeding strategies on crop diversity (Nightingale, 1996). It is also used to predict the amount of variability resulting from crosses (Souza and Sorrells, 1991). COPs can be used to measure heterosis in the offspring of crosses (Cox and Murphy, 1990) as well as to estimate the similarity of modern varieties (Souza *et al.*, 1994).

Several disadvantages have been noted with regard to the use of COPs as an indicator of genetic diversity. First, COPs only estimate the probability that alleles are identical by descent and provide no information on alleles that are identical by state (Nightingale, 1996). In other words, two alleles may be identical whether or not they share the same parentage. In calculating COPs based on pedigrees, one is necessarily limited by the pedigree as it is recorded; the ancestors positioned at the outermost leaves of the ancestral tree are typically understood as unrelated (Figure 2.1). Typically, the outermost leaves are as recorded by scientists in the first plant breeding program in which they were used, one to two hundred years ago. The assumption that they are unrelated may be incorrect from a biological or ecological perspective and probably biases the COP downwards (Cox *et al.*, 1985). Molecular methods now provide a means of testing the degree of relatedness of ancestors.

The algorithms employed to calculate COPs impose the assumption that each parent contributes equally to offspring, despite the effects of recurrent selection and random genetic drift. This assumption is unlikely to be valid, especially for qualitative traits such as disease resistance and height. Although the COP between a cultivar and a selection from it must clearly be less than one, its value is usually assigned somewhat arbitrarily by the researcher. The COP is thus unable to capture the effects of recurrent selection on allele frequencies in genotypes (Nightingale, 1996). Molecular methods offer techniques that can more accurately partition genetic contributions by ancestor, thereby improving the scientific basis of the algorithms used to calculate COPs.

Despite these disadvantages, an extensive animal breeding, crop science, and social science literature documents researchers' use of COPs. For examples of use in the crop science literature, see Cox, Murphy, and Rodgers (1986); Souza and Sorrells (1991); Murphy, Cox, and Rodgers (1986); van Beuningen (1993); Granet, Ludwig, and Melchinger (1994); van Hintum and Haalman (1994); Souza *et al.* (1994); Kim (1995); Plaschke, Ganal, and Roder (1995); Nightingale (1996); and del Toro (1996). Heisey *et al.* (1997) and Pray and Knudson (1994) have used COPs in economic analysis.

2.5. INTER-VARIETAL DIVERSITY AND INTRA-VARIETAL DIVERSITY

Crop diversity can also be differentiated by the amount of diversity found among crop populations ("inter-varietal diversity") as opposed to the amount of diversity found within a given crop population ("intra-varietal diversity"). Inter-varietal diversity and intra-varietal diversity are relevant concepts at various levels of analysis, including the farm household, the region, or a breeding program. Inter-varietal diversity at the household level often refers to the number of named varieties cultivated on the farm. Intra-varietal diversity at the household level could be measured using morphological or molecular techniques. At the regional level, inter-varietal diversity could be represented as spatial or temporal diversity, e.g., the number of named or otherwise-categorized varieties in a certain area or over a selected time period. In a breeding program, intra-varietal diversity might imply the number of genes conferring resistance to a particular disease that are found in one variety.

The concept of intra-varietal diversity takes into consideration the possibility that variation occurs within a given crop population or named variety. In the case of a self-pollinating crop such as wheat or rice, this variation may be the result of seed mixture or another human or natural activity that results in the reduction of seed purity. For self-pollinating crops, intra-varietal diversity is potentially most important in areas of crop diversity where seed is infrequently replaced (Meng, 1997). In contrast, for open-pollinated crops such as maize, diversity within a named variety is often the result of the deliberate replacement, exchange, or mixing of seed by farmers (Louette, 1994; Aguirre, 1997).

The nature of intra-varietal diversity requires the use of another means of characterizing the population besides the variety's name. For example, multivariate classification methods such as cluster analysis can be used with data on qualitative and quantitative characters to define the subgroup structure of populations and predict group membership (Franco et al., 1997). Because crop populations are often identified by name, exclusive reliance on names to determine diversity levels may underestimate or overestimate the actual amount of diversity. The presence of intra-varietal diversity can thus be easily overlooked in the absence of information on individuals within the population, such as physically observable characteristics or genetic markers.

Although intra-varietal diversity is most commonly associated with traditional varieties, it is not limited to them. This type of diversity has also been observed in advanced generations of improved varieties. Moreover, intra-varietal diversity and inter-varietal diversity are not mutually exclusive—both can be present at either the household or aggregate level. It is also possible for intra-varietal diversity to be high at the household level and low at a regional or provincial level, and vice versa.

2.6. INDICES

Given a decision on the most appropriate concept of diversity and the determination of the means to represent it (e.g., named varieties, morphological traits, molecular markers, pedigree, etc.), a mathematical construct, often in the form of an index, is calculated to summarize the relevant data for incorporation into the application of an economic model or approach. One of the primary advantages of using such an index is that the data can be condensed into a scalar measure for use in an analytical application. Numerous diversity indices exist, many of which have been developed for ecological applications (Magurran, 1991). Some are specific to certain types of data (e.g., qualitative versus quantitative), but most are based in some way on the measurement of pair-wise distances between selected characteristics. The choice of index is important in that it should be appropriate for the type of data available and also for the type of diversity that is being examined.

The way diversity is structured within and between populations also depends on the mating system of the crop (Hamrick and Godt, 1997). A number of different indices can be constructed with molecular data, each expressing a different type of genetic variation. The genetic variation in a single variety or a group of varieties, and particularly in landraces, can occur as a large number of alleles at a single locus, but it can also occur as variation in the frequency of alleles among the populations of the same or distinct varieties.

In addition to the simple count measures commonly used to reflect spatial diversity, other indices include the proportion of area planted to the single most popular variety, the top five varieties, or the top ten varieties. A similar index is compiled from the number of varieties accounting for a given percentage of cultivated area (Widawsky, 1996). Another measure, the Herfindahl index, borrowed from the economic literature on industrial organization, is defined as the sum of squared shares of total crop area planted to each unique variety (Pardey et al., 1996; Hartell, 1996). The Shannon–Wiener (also known as the Shannon) index and Simpson index are additional examples of indices commonly used to measure spatial diversity. Both of these indices are measures of proportional abundance; that is, they combine indicators of crop population richness as well as abundance (Magurran, 1991). The Simpson index is equal to one minus the Herfindahl index. Dominance indices, such as the Simpson index, are relevant particularly for inter-varietal diversity, because they provide a measure not only of the number of distinct populations planted, but also of their importance. The Shannon–Wiener index, originally used in information theory but commonly applied to evaluate species diversity in ecological communities, has been widely used in the agronomic literature to transform qualitative traits into a scalar measure that can be compared over sets of varieties (Spagnoletti Zeuli and Qualset, 1987; Jain et al., 1975).

Indices of temporal diversity, such as the average age and weighted (by area) average age of varieties grown by farmers have been proposed, used, and reviewed by Brennan and Byerlee (1991) and Brennan and Fox (1995). Weighting the age of a variety by its area distribution captures to some extent the effects of its diffusion pattern across space. An index could also be developed from the parameters that describe the shape of diffusion curves, using curves fitted to actual data on area distributions by variety.

The taxonomic tree has been used to measure diversity among biological species, and it can also be used to measure subspecies diversity. To identify which species to conserve, Weitzman (1992) proposed a distance measure that maximizes diversity among the surviving members of the set. Such a conservation goal would reflect the notion that the greater the distance among the members of the set, the less likely it is that the species will contain redundant characteristics. Solow, Polasky, and Broadus (1993) also discuss a measure that minimizes the distance between the surviving and extinct species. The conservation goal behind this measure is to preserve the most representative sample of current species as possible.

Both Weitzman (1992) and Solow, Polasky, and Broadus (1993) identify three fundamental properties that should be associated with any "sensible" index of diversity: (1) diversity should not be decreased by the addition of another species (monotonicity in species); (2) diversity should not be increased by the addition of a species that is identical to a species that is already in the set ("twinning"); and (3) diversity should not be decreased by an unambiguous increase in the distances between species (monotonicity in distance). The representation of distances among species or subspecies as a taxonomic tree, in which the distance measure is a scalar calculated from the total length of the branches of the tree, is consistent with these properties (Weitzman, 1992, 1993). Taxonomic trees can be calculated using morphological, genealogical, or genetic data. Solow, Polasky, and Broadus (1993) introduce the ecological concept that distance measures should account for the size of the set (species richness) as well as the distances among the members of the set. Theoretically, both the Weitzman and Solow–Polasky measures also permit weighting to express abundance of species or subspecies.

Hartell *et al.* (chapter 9) use the pair-wise coefficients of diversity among varieties and Ward's fusion strategy (a clustering method which is relatively insensitive to distributional assumptions) to generate a dendrogram and calculate the total length of the taxonomic tree for the sets of varieties grown each year in each district of Punjab, Pakistan. The same method was used in Smale and McBride (1996) for sets of the major varieties grown in regions of the developing world in 1990. The relative abundance of varieties was not captured in these calculations, and Hartell *et al.* used a separate index of spatial diversity in their analysis. Widawsky and Rozelle (chapter 10) develop and apply a modified version of the Solow–Polasky index, incorporating abundance through area distributions.

2.7. SUMMARY

This chapter has raised several key measurement issues of importance to economists studying the diversity of crop plants. First, there is no single most accurate or most appropriate indicator of diversity in crop genetic resources; there are many indicators, of which the most appropriate will depend on the objectives of the study. Second, all measurements are based on the ability of researchers or farmers to group traits, markers, individuals, or populations to derive an indicator of distance or dissimilarity among them. Finally, social scientists must remember that farmers may or may not perceive or be interested in the characteristics that are used to measure diversity. Consequently, in modeling and estimating the effects of farmers' decisions on crop diversity, the relationship between diversity indices and behavior is likely to be both indirect and difficult to establish empirically.

Note

1 Three additional factors may affect allele frequencies: (1) rare mutation events; (2) migration from other varieties; and (3) genetic drift, or the random process of allele loss that depends particularly on population size. The first two factors may act to increase polymorphism, although in general the importance of mutations is very small compared to the migration factor within a few cropping seasons. The third factor exerts pressures in the opposite direction.

References

Aguirre, A. 1997. *Análisis regional de la diversidad del maíz en el Sureste de Guanajuato*. Ph.D. thesis, Universidad Nacional Autónoma de México, Facultad de Ciencias, Mexico, D.F.

Apple, J. L. 1977. The theory of disease management. In J.G. Horsfall (ed.), *Plant Disease: An Advanced Treatise*. New York: Academic Press.

Begossi, A. 1996. Use of ecological methods in ethnobotany: Diversity indices. *Economic Botany* 50(3): 280–289.

Brennan, J. P., and D. Byerlee. 1991. The rate of crop varietal replacement on farms: Measures and empirical results for wheat. *Plant Varieties and Seeds* 4: 99–106.

Brennan, J. P., and P. Fox. 1995. *Impacts of CIMMYT Wheats in Australia: Evidence of International Research Spillover*. Economics Research Report No. 1/95. Wagga Wagga: NSW Agriculture.

Cox, T. S., Y. T. Kiang, M. B. Gorman, and D. M. Rodgers. 1985. Relationship between coefficient of parentage and genetic similarity indices in the soybean. *Crop Science* 25 (May–June): 529–532.

Cox, T. S. and J. P. Murphy. 1990. The effect of parental divergence on F_2 heterosis in winter wheat crosses. *Theoretical and Applied Genetics* 79: 241–250.

Cox, T. S., J. P. Murphy, and D. M. Rodgers. 1986. Changes in genetic diversity in red winter wheat regions of the United States. *Proceedings of the National Academy of Sciences (US)* 83: 5583–5586.

del Toro, E. 1996. Impacto del ensayo de rendimiento de selección elite de trigo harinero (ESWYT) y su diversidad genética. M.Sc. thesis, Colegio de Postgraduados, Montecillos, Mexico.

Doebley, J. F., M. M. Goodman, and C. W. Stuber. 1985. Isozyme variation in the races of maize from Mexico. *American Journal of Botany* 72(5): 629–639.

Dudley, J. 1994. Comparison of genetic distance estimators using molecular marker data. In American Society for Horticultural Science (ASHS) and Crop Science Society of America (CSSA) (eds.), *Analysis of Molecular Marker Data, Joint Plant Breeding Symposia Series*. Corvallis: ASHS, CSSA.

Duvick, D. 1984. Genetic diversity in major farm crops on the farm and in reserve. *Economic Botany* 38(2): 161–178.

Ford, E. B. 1940. Polymorphism and taxonomy. In J. S. Huxley (ed.), *New Systematics*. Oxford: Clarendon Press.

Franco, J., J. Crossa, J. Villaseñor, S. Taba, and S. A. Eberhart. 1997. Classifying Mexican maize accessions using hierarchical and density search methods. *Crop Science* 37: 972–980.

Gollin, D., and R. E. Evenson. 1990. Genetic resources and rice varietal improvement in India. Mimeo. New Haven: Yale University, Economic Growth Center.

Granet, A., W. F. Ludwig, and A. E. Melchinger. 1994. Relationships among European barley germplasm. II: Comparison of RFLP and pedigree data. *Crop Science* 24: 1199–1205.

Hamrick, J. L., and M. J. W. Godt. 1997. Allozyme diversity in cultivated crops. *Crop Science* 37: 26–30.

Hartell, J. 1996. The contribution of genetic resource diversity: The case of wheat productivity in the Punjab of Pakistan. M.Sc. thesis, University of Minnesota, St. Paul, Minnesota.

Heisey, P. W., and J. P. Brennan. 1991. An analytical model of farmers' demand for replacement seed. *American Journal of Agricultural Economics* 73: 1044–1052.

Heisey, P. W., M. Smale, D. Byerlee, and E. Souza. 1997. Wheat rusts and the costs of genetic diversity in the Punjab of Pakistan. *American Journal of Agricultural Economics* 79: 726–737.

Jain, S. K., C. O. Qualset, G. M. Bhatt, K. K. Wu. 1975. Geographical patterns of phenotypic diversity in a world collection of durum wheat. *Crop Science* 15: 700–704.

Kempthorne, O. 1969. *An Introduction to Genetic Statistics*. Ames: Iowa State University.

Kim, H. S. 1995. Genetic diversity among wheat germplasm pools with diverse geographical origins. Ph.D. thesis, Michigan State University, East Lansing, Michigan.

Louette, D. 1994. Gestion traditionnelle de variétés de maïs dans la réserve de la Biosphère de Manantlán (RBSM, états de Jalisco et Colima, Mexique) et conservation in situ des ressources génétiques de plantes cultivées. Ph.D. thesis, École Nationale Supérieure Agronomique de Montpellier, Montpellier, France.

Louette, D., A. Charrier, and J. Berthaud. 1997. *In situ* conservation of maize in Mexico: Genetic diversity and maize seed management in a traditional community. *Economic Botany* 51(1): 20–38.

Magurran, A. 1991. *Ecological Diversity and Its Measurement*. Princeton: Princeton University Press.

Malecot, G. 1948. *Les Mathematiques de l'Hérédité*. Paris: Masson.

Meng, E. C. H. 1997. Land allocation decisions and *in situ* conservation of crop genetic resources: The case of wheat landraces in Turkey. Ph.D. thesis, University of California, Davis, California.

Murphy, J. P., T. S. Cox, T. S., and D. M. Rodgers. 1986. Cluster analysis of red winter wheat cultivars based upon coefficients of parentage. *Crop Science* 26: 672–676.

Nightingale, K. J. 1996. Trends in genetic diversity of spring wheat in the developing world. M.Sc. thesis, Michigan State University, East Lansing, Michigan.

Pardey, P. G., J. M. Alston, J. E. Christian, and S. Fan. 1996. *Summary of a Productive Partnership: The Benefits from U.S. Participation in the CGIAR*. EPTD Discussion Paper No. 18. Washington, DC: International Food Policy Research Institute (IFPRI) and University of California, Davis.

Plaschke, J., M. W. Ganal, and M. S. Roder. 1995. Detection of genetic diversity in closely related bread wheat using microsatellite markers. *Theoretical and Applied Genetics* 91: 1001–1007.

Plucknett, D. L., and N. J. H. Smith. 1986. Sustaining agricultural yields. *BioScience* 36: 40–45.

Pray, C. E., and M. Knudson. 1994. Impact of intellectual property rights on genetic diversity: The case of US wheat. *Contemporary Economic Policy* 12: 102–112.

Smale, M., and T. McBride. 1996. Understanding global trends in the use of wheat diversity and international flows of wheat genetic resources. Part I of *CIMMYT 1995/96 World Wheat Facts and Trends: Understanding Global Trends in the Use of Wheat Diversity and International Flows of Wheat Genetic Resources*. Mexico, D. F.: International Maize and Wheat Improvement Center (CIMMYT).

Solow, A., S. Polasky, and J. Broadus. 1993. On the measurement of biological diversity. *Journal of Environmental Economics and Management* 24: 60–68.

Souza, E., P. Fox, D. Byerlee, and B. Skovmand. 1994. Spring wheat diversity in irrigated areas of two developing countries. *Crop Science* 34: 774–783.

Souza, E., and M. E. Sorrells. 1989. Pedigree analysis of North American oat cultivars released from 1951 to 1985. *Crop Science* 29 : 595–601.

Souza, E., and M. E. Sorrells. 1991. Prediction of progeny variation in oat from parental genetic relationships. *Theoretical and Applied Genetics* 82: 233–241.

Spagnoletti Zeuli, P. L., and C. O. Qualset. 1987. Geographical diversity for quantitative spike characters in a world collection of durum wheat. *Crop Science* 27: 235–241.

van Beuningen, L. T. 1993. Genetic diversity among North American spring wheat cultivars as determined from genealogy and morphology. Ph.D. thesis, University of Minnesota, St. Paul, Minnesota

van Hintum, T. J. L., and D. Haalman. 1994. Pedigree analysis for composing a core collection of modern cultivars, with examples from barley (*Hordeum vulgare* s. lat.). *Theoretical and Applied Genetics* 88: 70–74.

Weitzman, M. L. 1992. On diversity. *Quarterly Journal of Economics* 107: 363–397.

Weitzman, M. L. 1993. What to preserve? An application of diversity theory to crane conservation. *Quarterly Journal of Economics* 108: 1557–1583.

Widawsky, D. A. 1996. Rice yields, production variability, and the war against pests: An empirical investigation of pesticides, host-plant resistance, and varietal diversity in Eastern China. Ph.D. thesis, Stanford University, Palo Alto, California.

Wright, S. 1922. Coefficients of inbreeding and relationship. *American Naturalist* 56: 330–338.

Valuation of Crop Genetic Resources *Ex Situ*

3 THE COST OF CONSERVING MAIZE AND WHEAT GENETIC RESOURCES *EX SITU*

P. G. Pardey, B. Skovmand, S. Taba,
M. E. Van Dusen, and B. D. Wright[1]

3.1. INTRODUCTION

Although the technical performance attained in long-term storage of crop germplasm has improved dramatically over the past several decades, key management questions remain to be addressed (Wright, 1997). These include issues related to the size of gene banks, or the number and type of genetic materials that should be conserved, as well as their utilization (Frankel, Brown, and Burdon, 1995). A comprehensive and accurate economic evaluation of an *ex situ* conservation program would weigh the benefits against the costs to assess net benefits, requiring the estimation of the marginal benefits of conserving each type of genetic material. However, estimating benefits is methodologically and empirically challenging, for several reasons.

First, attributing the appropriate part of the gains realized through plant breeding to the use of conserved germplasm is a difficult, if not intractable, inferential problem (Pardey *et al.*, 1996a, 1996b). Second, modern gene bank facilities are so new that insufficient time has elapsed to establish a time series of realized gains attributable to their establishment. Third, in addition to the use value on which such estimates are typically based, conservation of genetic diversity in crop germplasm yields an option value for responding to currently unidentified future demands. To some people at least, conservation also yields existence value; people report they are better off simply for knowing crop diversity is not being lost. Although economic methodologies are available to assess option and existence value, estimates developed on the basis of such methodologies are likely to be imprecise.

This study is based on a more tractable approach: if the total and marginal costs of operating a gene bank appear less than any reasonable lower-bound estimate of the corresponding benefits, then a complete assessment of benefits is unnecessary. If there is any doubt that the lower-bound estimate of benefits exceeds costs, the benefits of operating the gene bank also warrant further investigation. Estimating costs, as compared with estimating benefits, involves assembling historical data on gene bank operations, which are usually found in financial records and supplemented by interviews with technical and accounting staff.

The example used here is the gene bank at CIMMYT, where both maize and wheat genetic resources are conserved in the same facility. CIMMYT's collections of maize and wheat are among the largest in the world, with 123,000 wheat accessions and 17,000 maize accessions held in trust in 1995 under agreement between the Consultative Group on International Agricultural Research (CGIAR) and the Food and Agriculture Organization of the United Nations (FAO 1996) (Table 3.1). These accessions include cultivated maize and wheat varieties, landraces, breeding materials, wild grasses, and crop relatives provided by collaborating institutions around the world.

The next section summarizes an economic approach to estimating costs of gene bank operations, which is based on production theory. This is followed by information on the cost categories that need to be considered in this type of economic analysis, including the storage of the germplasm, regeneration of accessions, data management, maintenance of seed health, and shipping of materials to collaborating institutions. Cost estimates are then interpreted, with reference to studies conducted for different crops. Conclusions and implications for further research are discussed in the final section. Additional details, including a brief history of the wheat and maize collections held at CIMMYT, can be found in Pardey et al. (1998).

Table 3.1. Number of Accessions in CIMMYT Gene Bank, 1995

Crop/type	Number of accessions
Wheat collection	
Bread wheat	71,171
Durum wheat	15,490
Triticale	15,200
Barley	9,084
Rye	202
Primitive and wild	11,794
Total	122,941
Maize collection	
Zea	17,000

Source: FAO (1996) and CIMMYT data files.

3.2. THE ECONOMICS OF OPERATING A GENE BANK

Production economics provides a basic framework to use in developing cost estimates for the operation of a gene bank. Inputs such as labor, land, buildings, energy, and acquired seeds are used to produce stored seeds as well as the information that accompanies them, and to distribute seeds over the years to scientists at CIMMYT and collaborating national and international institutions.

Aside from generating accurate estimates of total annual costs of conserving wheat and maize genetic resources, a principal interest of this study has been to investigate economic issues of size, scale, and scope. Disaggregating total costs into variable and fixed (durable) cost components to enable calculation of average and marginal costs also reveals some information regarding these issues, though based on a single year's estimate. Average annual storage costs can be calculated as the total costs of storage (in a given year) divided by the number of accessions in a collection. Marginal costs are the additional costs involved in conserving another accession.

Reviews and discussions of the concepts of economies of size, scale, and scope include those found in Bailey and Friedlaender (1982) and Baumol, Panzar, and Willig (1988). Here, "economies of size or scale," loosely speaking, refer to reductions in the unit costs of producing stored seed that are associated with increases in the size of the operation. Economies of size or scale are often a consequence of specialization or more efficient allocation of labor inputs. For example, larger operations might provide the opportunity to hire technicians to sort and classify seed, re-allocating the relatively costly time of geneticists and agronomists from these tasks to management and evaluation. Economies of size are exploited when the gene bank facility is large enough to have (and efficiently use) lumpy, fixed factors of production and other variable inputs. Fixed factors of production might include physical infrastructure as well as scientific expertise or human capital.

By comparison, economies of scope mean that it is cheaper to simultaneously produce multiple outputs than to produce one output, as a result of either technical or cost complementarities. Unit costs of producing one output decline with production of the other output (Leathers, 1991). The classical example of technical complementarity in production is the case of joint products, such as wool and mutton. Economies of scope can also occur when fixed costs can be shared between two or more production processes or when a cost constraint imposes an interrelationship between the cost structures of the production processes. Thus, consolidating the wheat and maize collection in a shared facility may offer the prospects of significant cost savings compared with maintaining each crop collection in a separate facility.

Using a production economics framework offers the potential to analyze other aspects of gene bank operations, such as those related to technical change and allocative efficiencies. As the relative prices of inputs such as labor, capital, and chemicals change, so should the mix of those inputs in the storage and distribution of seed. The sensitivity of the mix of inputs to the changes in relative prices is in turn dependent on the degree of substitutability and complementarity of the respective inputs, aspects that relate to the nature of the technologies used to produce stored seed. An increase in the price of labor over time, for instance, ought to spur a substitution of other inputs, such as electronic data processing, for labor, with consequences for the change in cost shares of the respective inputs in the total costs of the gene bank operations. The technology of gene bank operations over time will also affect the optimal amount and mix of inputs used in the longer run. Moreover, as suggested by the induced innovation model of technical change (Hayami and Ruttan, 1985), the direction and nature of the technical change may itself be driven by shifts in relative prices, which tends to reinforce in the longer run the magnitude and direction of the shorter-term shifts in input mix initiated by the price changes.

Knowing the costs of operating the gene bank is essential to this and other types of economic analyses related to genetic resource conservation and utilization. For example, the cost of storing an accession is required to more fully value marginal accessions through the use of search models like those proposed in this book. From the standpoint of a single search for a trait of economic value among gene bank accessions, the cost of operating the gene bank is fixed, and only the variable cost of searching for the trait is relevant to decisions regarding the optimal number of accessions to search (see chapters 4 and 5). To value a marginal accession, or to determine the optimal size of a collection, however, repeated searches over single and multiple traits must be considered and operational costs included.

The issue of information and characterization relates to efficient search as well as the costs of operating the gene bank. If accessions have been characterized with respect to all potential traits, searching accessions is analogous to analysis of a bibliographic search in a library (Cooper and DeWath, 1976). Although scientists commonly state that inadequate characterization inhibits the utilization of accessions, current research raises questions about this claim. Koo and Wright (1998) have addressed the economics of searching for disease-resistance traits in gene bank accessions and transferring those traits to high-yielding, elite cultivars. They find that for sufficiently rare diseases, pre-characterization of a resistance trait does not make economic sense. The greatest value from pre-characterization, all else being equal, accrues with respect to resistance traits from diseases that occur with intermediate frequency. The cost decreases that may accrue from advances in biotechnology encourage pre-characterization. For example, as technologies for detecting clones become cheaper and more effective, they can be used to screen accessions before they enter the gene bank, substituting for land and labor used in maintaining field gene banks, and in the longer run substituting for space in long-term seed storage facilities.

3.3. COSTS OF OPERATING THE CIMMYT GENE BANK

3.3.1. Overview of Cost Categories

Gene bank expenses were categorized into variable and durable costs as a basis for distinguishing between average and marginal costs. Variable costs involved current expenses that were often sensitive to the scale of operations, such as the labor, fertilizer, and other chemical costs involved in regenerating gene bank accessions. Durable costs involved costs that were incurred on a periodic (greater than one year) basis, such as the costs of the gene bank facility and the physical plant required to refrigerate it.

3.3.2. The New CIMMYT Facility

In 1996, a new gene bank facility was completed at CIMMYT, and for the first time the maize and wheat collections were consolidated into a single facility with advanced technology for medium- and long-term storage. The main structure of the new gene bank facility consists of a two-story, fortified-concrete bunker, built to withstand most conceivable natural or manmade disasters. The climate is controlled to precise temperature and humidity specifications and the facility is equipped with alarms, security measures, and a backup power supply. The upper (ground) level of the storage rooms houses the active collection, stored at just below freezing (–1°C) and 30% relative humidity. This constitutes the "working" part of the bank, from which seed requests from scientists are filled. The lower (below-ground) level consists of the base collection stored at –18°C. This area is used primarily for long-term storage. The seeds are stored on movable shelves to optimize use of the available space.

The size of the seeds themselves is a source of distinction between the maize and wheat holdings that has significant management and cost implications. A sample of wheat is 250 g (about 7,000 seeds), whereas the working sample of maize is 3 kg (6,000–10,000 seeds). Wheat accessions are stored in aluminum laminated envelopes the size of a one-pound bag of coffee (about 500 g), while maize accessions are stored in one gallon plastic containers (about 4 L). The new facility allocates 240 m³ of both medium- and long-term storage space to each program, sufficient to store 390,000 wheat accessions and 67,000 maize accessions in the base collection. If present rates of growth in the size of the respective collections persist, it will take 53 years to fill the space allocated to wheat and 50 years to fill the space set aside for maize.

3.3.3. Durable Input Costs

Construction costs for the new gene bank facility are shown in Table 3.2. Complementing the new storage facility are various rooms for sorting and packing seeds destined for storage or for shipment, a drying room, warehouses, offices, and a laboratory shared by the staff of the maize and wheat gene banks. The construction of much of this ancillary space involved renovating existing facilities whose cost is reported under the category of overhead costs in the consolidated budget of CIMMYT. Part of the capital costs associated with the gene bank structure involves the use of backup power generation units in CIMMYT's physical plant facility. Rather than treat these costs as part of the capital costs of the gene bank, we included them in a separate set of calculations related to the costs of refrigerating and dehumidifying the storage facility.

To estimate a so-called "user cost" (in this instance the annualized, equivalent cost) of outlays on lumpy capital items such as buildings and equipment, an appropriate and often convenient method is to treat commercial rental rates of the relevant capital items as an estimate of the annual user cost of capital. In the absence of relevant rental rates, we directly estimated annual capital costs based on information about the purchase price of each capital item, assumptions about its service life, and the real rate of interest.[2]

The building in which the seed holdings are stored is deemed impervious to destruction and is likely to have a long service life; we took it to be 40 years (an estimate that is also in line with general CGIAR depreciation guidelines). The service life of the climate-control machinery was assumed to be 15 years (this equipment is

Table 3.2. Construction Costs of the New CIMMYT Gene Bank

Cost category	Expenditures (1995 US$)	Share of total costs (%)
Construction (materials, labor, etc.)	351,034	34.3
Electrician (Mexico)	76,611	7.5
Electrician (US)	75,144	7.3
Thermal panels	93,237	9.1
Refrigeration equipment	102,914	10.0
Shelving	225,000	22.0
Elevator (US)	8,400	0.8
Elevator (Mexico)	3,526	0.3
Project assistant	20,167	2.0
Project manager	51,280	5.0
Legal fees	3,105	0.3
Soil mechanics	3,755	0.4
Miscellaneous structures	655	0.1
Laboratory (germination, seed counting, etc.)	2,006	0.2
Reimbursements	2,099	0.2
Reimbursements	978	0.1
Insurance	1,602	0.2
Paperwork	1,739	0.2
Miscellaneous	866	0.1
Total	1,024,118	100.0

Source: Unpublished CIMMYT documentation prepared by GAO Consultants.

running almost constantly and subject to wear and tear). The structures and equipment were taken to have zero salvage value at the end of their working lives, and we assumed a real discount rate of 4%.[3]

On the presumption that the germplasm is to be held in perpetuity, and given the finite service lives of the capital items in this study, we estimated the annual user cost of capital based on the notion that the capital items were repurchased at regular intervals (40 years for the building, 15 years for the climate-control equipment).[4] The algebra for this problem is spelled out in an appendix to this chapter. The annual user costs estimated here represent the real (adjusted for inflation) funds that need to be set aside to cover the initial capital outlays and their subsequent replacement into perpetuity. User costs for a representative year are given in Table 3.3.

Maintaining the storage areas in the gene bank at a precise, stable, low temperature–low moisture (i.e., low relative humidity) regime is a costly exercise. The current costs of controlling the climate in the facility comprise the electricity costs involved in running the compressors, dehumidifiers, and fans, the costs incurred in maintaining this equipment, and the related costs of maintaining and running the emergency power plant (Table 3.4). Allocating these costs to the germplasm facility is difficult as they represent only part of the overall costs involved in operating the physical plant. In calculating the costs, we included the energy required to maintain the gene bank at its specified climate characteristics, and also estimated the costs associated with a stylized (but plausible) schedule of maintenance on the climate-control equipment and the backup power generation unit (further details can be found in Pardey *et al.*, 1998). Both sets of calculations were developed based on 1996–1997 figures.

3.3.4. Monitoring and Regenerating Seed

El Batán, Mexico, where the maize and wheat gene banks are located, is in a seasonally dry and wet tropical highland climate, which is favorable to seed conservation as

Table 3.3. Annualized Capital and Recurrent Costs for the CIMMYT Gene Bank Facility

Cost category	Cost (US$)
Annualized capital costs[a]	53,692
Buildings	44,752
Refrigeration equipment	8,900
Recurrent costs	46,770
Total annual costs	100,462

[a] Calculated according to the method described in the appendix to this chapter.

Table 3.4. Operational and Maintenance Costs for Gene Bank Climate Control and Power Units

	Kilowatts	Hours per year	Kilowatt hours	US dollars per year
Compressors				
Active	11.19	2,920	32,675	2,178
Base	22.38	2,920	65,350	4,357
Heater	15.00	1,460	21,900	1,460
Heat pump	0.75	2,920	2,178	145
Fans	1.68	2,920	4,901	327
Subtotal				9,927
Dryers				
Storage	54.87	1,460	80,110	12,818
Maize	24.01	1,460	35,048	7,510
Wheat	27.43	1,460	40,055	8,592
Subtotal				28,920
Maintenance			**Work months**	
Refrigeration	–	–	9.6	2,462
Cleaning[a]	–	–	24.0	4,615
Electrical	–	–	3.3	846
Subtotal	–	–	–	7,923
Total	–	–	–	46,770

Note: Energy costs estimated using "equipment requirements" method, which involved estimating energy consumption required to perform various functions according to specifications for an average year. Labor used for estimating maintenance costs (i.e., technician's time) obtained from CIMMYT engineering office.

[a] Includes cost of cleaning gene bank storage and seed processing areas.

compared with year-round tropical and humid conditions. Nonetheless, as seeds age they gradually lose their viability in storage. The monitoring and regeneration procedures followed by CIMMYT begin with a test of viability five to ten years after the seed is first introduced into the gene bank or was last regenerated. A sample of the seed from each accession is placed in a germinator for several days and checked to determine its germination rate. If that rate falls below 90% of the respective baseline rates for wheat and maize, the accession undergoes a cycle of regeneration. A large share of the costs in assessing viability involve the costs of the labor used to implement the tests. Some additional overhead costs, such as those of maintaining a suitable laboratory with germination chambers, must also be included. It is also the case that some accessions must be replenished as stocks are depleted through the distribution of samples on request to scientists.

A principal challenge in managing the regeneration of an *ex situ* collection is the potential for random genetic drift in the collection. In population genetics, "genetic drift" generally refers to the loss of alleles that results from small population sizes (Falconer, 1981). In the regeneration of gene bank accessions, it is caused by drawing progressively smaller numbers of individuals from the initial population, which is itself a sample from a larger population (Crossa *et al.*, 1994). One of the significant advantages of the new gene bank is that the long-term storage facility is held at −18°C, which should enable seed to remain viable for over 50 years. Even if the sample size is within generally accepted guidelines, repeated draws from an especially heterogeneous sample may shift the genetic structure of the accession away from the original sample. Moreover, random genetic drift may be exacerbated if samples are regenerated under conditions of soil, chemical inputs, or daylight that differ markedly from the native ecology, but this is generally more of a concern when regenerating wild relatives than more advanced breeding lines and improved varieties.

As a general rule, the rates of genetic drift are much less for a self-pollinating crop like wheat than for an open-pollinating crop like maize. The different regeneration procedures for maize and wheat reflect these different crop characteristics. The CIMMYT wheat bank chooses to regenerate an entire accession, replacing both the active and base collections when the viability falls below threshold levels. The maize bank opts to sample from the base collection to restore the active collection. The objective of both of these methods is to maintain in the active collection a genetic makeup that resembles as closely as possible that of the original maize holdings.

Wheat accessions are first regenerated for seed in a screenhouse at El Batán, arranged in one- to three-hill plots that include at least 25 plants per plot. The screenhouse facility enables regeneration to proceed on a year-round basis, with up to three cycles per year under controlled and protective conditions. Seed samples prepared in the screenhouse are then flown to Mexicali, in the state of Baja California Norte, where they are sown out in 1 m rows to scale up the size of the sample to 500 g. The peak labor requirements in the regeneration process occur at harvest, when field books in which the physiological traits for each accession are recorded must be completed.

Most of CIMMYT's maize accessions that are derived from tropical maize-growing areas of low and intermediate elevation are regenerated at Tlaltizapán, whereas tropical highland germplasm is regenerated at El Batán. Maize accessions that are not adapted to CIMMYT stations, such as those from the Andean region, should be regenerated in source countries. Currently 2.5 ha/yr are made available for regenerating maize accessions, sufficient for 5,000 rows of 5 m. Each attempt at regeneration may not yield a sufficient quantity of seed, determined to be 100 usable ears. The first round of regeneration gives acceptable results only 60% of the time; as a consequence, a significant share of the 5,000 rows in each regeneration cycle involves a second effort to regenerate the 40% of failed samples carried over from the first round.

Table 3.5 reports typical input costs required to prepare 1 ha for planting at El Batán. Similar calculations were also performed for the Tlaltizapán site. A number of inputs such as irrigation, agrochemicals (including fertilizers), and management time vary according to seasonal and other factors, but our cost estimates provide a benchmark that is useful for our purposes. Given that many of these costs are not explicitly itemized as such in CIMMYT's accounting system, we first estimated the typical quantity of

Table 3.5. Summary of Baseline Field Costs for Regenerating Germplasm at El Batán

Field preparation and planting activity	Units (number)	Cost/ha (MX$)	Cost/ha (US$)
Initial land preparation			
Clearing	2 h	200	25.64
Rake	1 pass	150	19.23
Ripping	1 pass	300	38.46
Rake	1 pass	150	19.23
Fertilizer	163 kg	326	41.79
Fertilizer	163 kg	326	41.79
Rows	1 pass	150	19.23
Fertilizer	87 kg	174	22.31
Incorporating fertilizer	3 passes	450	57.69
Making beds	1 pass	150	19.23
Sowing		300	38.46
Labor	2 d	100	12.82
Irrigation	12 h	120	15.38
Labor	6 d	300	38.46
Pest control (initial application)			
Brominal	2 L	300	38.46
Topik	0.25 L	850	108.97
Estarene	1 L	360	46.15
Basgaran	4 L	310	39.74
Lorsban	1 L	240	30.77
Post-seeding field expenses			
Cultivate	2 passes	300	38.46
Irrigation	3 applications	360	46.15
Labor		600	76.92
Pest control (application at harvest)			
Folicur	1 L	407	52.18
Tilt	1 L	450	57.69
Labor	20 applicators	1,000	128.21
Total		8,373	1,073.46

Source: Authors' calculations, based on data provided by M. Magallanes and A. López at CIMMYT.

each of the inputs required to prepare, plant, and harvest 1 ha, priced each item accordingly, and then derived the corresponding cost. A shadow rental rate was the basis of the estimates for the user cost of land included in the regeneration costs.

Based on the cost elements detailed in Table 3.5, the benchmark field-related cost to plant, tend, and harvest a regenerating crop in El Batán is US$ 1,073/ha and US$ 1,173/ha in Tlaltizapán. We used these benchmark, per-hectare field costs to estimate the overall costs involved in regenerating an accession of wheat and one of maize, adjusting these benchmark costs to reflect differentials that arise due to differences in the seed density, volume, and reproductive characteristics of each crop. For instance, it takes at least 60 m^2 to regenerate an accession of maize, whereas an accession of wheat typically requires only 0.75 m^2. The labor costs for maize are much higher than for wheat because of the hand pollination required for each plant. Regenerating maize also involves additional costs associated with the glassine and pollination bags that are used to control pollination. For wheat the costs include the plots at Mexicali and El Batán and an annualized estimate of the user cost of capital for a screenhouse. Table 3.6 summarizes the total and average costs for regenerating a typical accession of each crop. The difference in regeneration costs between the two crops is salient.

3.3.5. Processing New Seed Accessions

Processing a new accession is much like regenerating an existing accession but involves certain additional requirements. First the seed is inspected thoroughly before and after regeneration to screen for any known or suspected seed health problems. The first regeneration is performed on special quarantine plots that maintain stringent pest-control procedures. A number of additional observation, characterization, and data entry activities must be performed before an accession is finally added to the collection.

It typically takes far more time to manually clean, sort, and inspect maize seeds than it does wheat; each maize accession must be sorted individually by hand to remove broken or diseased seed. Although wheat seeds are intrinsically easier to handle, they require comparatively more attention to aspects of seed health. Both maize and wheat accessions require a similar amount of labor to record relevant data in field books, but the higher planting density for wheat affords it some time savings compared with maize. Both wheat and maize are stored at CIMMYT headquarters in two containers; one goes to the active collection, and the other is kept for long-term storage in the base collection. Wheat is stored in laminated foil bags, at a cost of US$ 0.11 each; the maize accessions are sealed in plastic buckets costing US$ 2.40 each. In addition, a sample of

Table 3.6. Total and Average Regeneration Costs

Crop	Regeneration costs (US$)				Number of accessions regenerated	Average cost (US$/accession)
	Field	Labor	Supplies	Total		
Wheat	15,027	11,863	–	26,890	22,000	1.22
Maize	9,094	31,751	12,000	52,845	650	81.30

Source: Authors' calculations.

each accession (10 g of wheat seed and 1.5 kg of maize) is prepared for backup storage in the United States National Seed Storage Laboratory in Colorado. A further 10 g sample of each wheat accession is prepared for backup storage at the International Center for Agricultural Research in the Dry Areas (ICARDA), Aleppo, Syria.

Before they are placed in storage, all seeds are dried to reduce the relative humidity after harvesting and cleaning. The maize bank uses two steps of seed-drying operations before a seed lot is deposited at the bank. After harvest from the regeneration field plots, the ears are dried in an air-forced drying room at 35–40°C to reduce the seed moisture to 12–15%. They are then shelled, and balanced bulk samples are made. The balanced bulk samples are transported to a cool drying room in the bank where the seed moisture is equilibrated to the drying room conditions over a period of two to three months, until moisture content reaches 6–8%. The maize drying room can service 2 t of seed at a time.

The 2 t capacity of the old drying room at the wheat gene bank was a significant bottleneck to its operations. When combined with an average sample size of 0.5 kg and a drying time of three months, this meant that 16,000 accessions could be dried per annum. A new dryer (installed during the course of this study) has the same capacity as the current piece of equipment but will reduce the drying time to two months, increasing the throughput to 24,000 accessions per year. The costs of operating and maintaining the dryers are included in Table 3.6, along with the annualized user costs associated with the purchase of the dryer.

3.3.6. Seed Health

The general operation of a well-managed seed bank involves periodic checking for ambient (air-borne) spores, monitoring the cleanliness of the machinery used in processing the seed, and precautionary measures to eliminate possible vectors, which at CIMMYT involves the daily washing, with bleach, of all walls and floors in areas where seeds are processed. The importance of issues relating to seed health, quarantine, and monitoring is illustrated by the infestation of Karnal bunt in the regeneration field at Ciudad Obregón, Sonora, which was used before 1987 by the CIMMYT wheat bank. Karnal bunt is not a particularly virulent or economically important disease for wheat (Fuentes–Davila, 1996), but its presence does limit the acceptability of seed by numerous national quarantine agencies.

Wheat seeds are now multiplied in clean plots at El Batán, checked for spores in bulked samples after being disinfected with chlorine, regenerated at Mexicali, and shipped back to El Batán in sealed containers. To eliminate Karnal bunt teliospores, a Jacuzzi-like system for cleaning seeds was developed. The seeds are placed in plastic baskets with metal mesh bottoms and suspended for three minutes in a chlorine bath while agitated by air bubbles. This system has enabled the wheat gene bank and international nurseries system (by which seed samples are shipped to numerous countries for evaluation in comparative trials) to continue operating effectively. When shipping seed samples internationally, there are a number of additional, and increasing, costs involved in dealing with a more and more complex set of seed health regulations and obtaining phytosanitary certification.

The Karnal bunt infestation provides some interesting perspectives on various aspects of germplasm management. Notably, the natural disaster threat to the wheat collection (to date) has not involved an earthquake or a flood, but rather a microorganism. The costs of dealing with the infestation involve additional regeneration costs, specialized shipping procedures, increased chemical applications, and increased costs of monitoring seed health.

3.3.7. Seed Distribution

There are various reasons why germplasm is shipped from the germplasm bank. The most basic reason is in response to an individual request from a scientist or collaborating institution; such requests are usually made directly to the bank manager. Seed is also sent to other gene bank facilities, often in the context of CIMMYT's work with collaborators in national agricultural research institutions. An example of this type of exchange is the Latin American Maize Project (LAMP). In the first phase of the project, more than 12,000 accessions from 13 countries were evaluated, and workshops on cooperative regeneration were implemented. In the second phase of the project, a core subset chosen to represent the phenotypic diversity of maize landrace collections in Latin America will be further characterized by molecular fingerprinting and by breeding crosses.

The cost of responding to seed requests includes determining the choice of accessions to be sent, assembly, treatment, and packaging of the chosen samples, as well as associated shipping costs. An important aspect of shipping seeds relates to the phytosanitary certification process, an increasingly time-intensive (and therefore costly) undertaking that draws on the time of the head of the seed health unit, the staff that prepare the necessary documentation, as well as the Mexican government's certification charges. Furthermore, CIMMYT seed must be accompanied by a material transfer agreement designed to protect the intellectual and other property claims surrounding the seed. Table 3.7 summarizes the shipments made from the gene bank since 1987 for maize and 1992 for wheat.

3.3.8. Data and Information Management

Fundamental to the gene bank is the management of the information that describes each accession. It is difficult to separate data and information management for the gene banks from the operations of the CIMMYT breeding program. To introduce new materials to the bank, passport data (detailing the source and origin of the seed) needs to be checked to avoid duplication with materials that have already been stored. If the incoming sample is unique, a sufficient amount of seed is required for storage in the base and active collections. Regeneration data and seed management data, such as initial germination percentage, seed traits, and seed moisture, are also needed. The routine operations of the gene bank therefore include the processing of field book observations collected as part of a trial conducted when the accession is new to the collection and at all subsequent regenerations, as well as the maintenance of a database that tracks the storage location, time in storage, seed viability history, and stock levels of each accession.

The wheat gene bank internally contracts for the services of a data entry team and is currently in the process of digitizing all of its old field books. The maize bank commits 1.75 years of its own staff to managing the regeneration information. These data are geared not only to the internal management needs of the CIMMYT gene bank but are also made available to others on demand. Catalogs in the form of CD ROM or web-based databases are gradually replacing printed publications.

The gene bank management systems are part of a broader effort at CIMMYT to improve the information base concerning the Center's extensive maize and wheat holdings. The past three years have seen the creation of the Genetic Resources Information Package (GRIP) with a combination of Australian and CIMMYT core funding, and incorporation into the CGIAR Systemwide Information Network for Genetic Resources (SINGER) funded with outside funds. The Wheat Gene Bank Management System recently has been incorporated into the International Wheat Information System (IWIS), a computer database system that integrates information from nursery trials to pedigree information and is able to trace genealogies of advanced breeding lines.

The data management system for the maize germplasm bank was initially developed in the late 1980s and has acquired new functions in order to process passport data, information on regeneration, evaluation, seed monitoring, and shipment. CD ROMs were published with passport information in 1988, 1992, and 1995 (as a part of the LAMP project) and in 1997, through SINGER. These include data for all of the maize bank accessions, including the original collection and the regenerations from the ongoing LAMP collaboration.

Table 3.7. Shipments from the CIMMYT Gene Bank

| | | Rest of the world | | | |
| | | Developing | Developed | | |
Crop/year	CIMMYT	countries	countries	Total	Total
Maize					
1987	2,400	1,667	447	2,114	4,514
1988	4,341	1,489	587	2,076	6,417
1989	5,093	1,238	1,378	2,616	7,709
1990	3,450	1,103	687	2,090	5,540
1991	2,231	508	117	625	2,856
1992	1,970	536	710	1,246	3,216
1993	3,740	818	1,813	2,631	6,371
1994	3,039	717	637	1,354	4,393
1995	2,542	264	532	796	3,338
1996	2,776	803	106	909	3,685
Wheat					
1992	2,278	561	115	676	2,954
1993	6,333	584	1,160	1,744	8,077
1994	1,026	3,793	703	4,696	5,722
1995	2,944	229	101	330	3,274
1996	12,890	133	1,200	1,333	14,223

3.4. INTERPRETING COST CALCULATIONS

The CIMMYT gene bank facility is large and has substantial excess capacity. A plausible assumption to make given unused capacity of the storage facility is that the CIMMYT maize and wheat gene banks are presently operating in the declining portion of the average cost curve. In production economics, this implies that the marginal cost curve lies below the average cost curve; thus average annual costs can be taken as an upper-bound approximation to the marginal costs of storage.

The average costs for the maize and wheat gene banks are shown in Table 3.8, with one-half of the total construction cost allocated to each crop. The average costs are shown for both the current number of holdings and for the predicted capacity of the bank for both maize and wheat. Excess capacity means that the average costs per accession for wheat and maize (US$ 0.41 and US$ 2.95, respectively) are more than three times what they would be at full capacity (US$ 0.13 and US$ 0.74).

This does not necessarily mean that the facility is too large: many of the capital costs are probably insensitive to size, and the ratio of volume to surface area declines with size, so refrigeration costs increase less than proportionally to the size of the facility. Furthermore, the annualized cost of storage *per se* is only a fraction (about 7% for wheat and maize) of the cost of the gene bank operations. Regeneration is a major element of cost (compare Tables 3.6 and 3.8), in terms of average cost per accession, and is much more costly in maize (US$ 81.30) than in wheat (US$ 1.22), reflecting differences in the growth habits of the two crops and the costs of methods used to limit genetic drift in the highly heterogeneous maize accessions. The overall average variable costs are US$ 0.36 for wheat, US$ 5.83 for maize, and the average durable inputs (or fixed) costs are US$ 1.46 and US$ 9.44, respectively, giving average total costs of US$ 1.82 and US$ 15.27 per year for wheat and maize accessions (Table 3.9).

Several key issues are raised by these cost estimates. First, the average variable costs are much lower than average total costs for both grains. The capital-intensive storage facility economizes on cooling costs and facilitates long time periods between

Table 3.8. Average Storage Costs

Annualized costs	Average cost at current capacity (US$/accession)		Average cost at full capacity (US$/accession)	
	Maize (17,000)[a]	Wheat (123,000)[a]	Maize (67,000)[b]	Wheat (390,000)[b]
Costs of durable inputs	1.58	0.22	0.40	0.07
Building	1.32	0.18	0.33	0.06
Refrigeration	0.26	0.04	0.07	0.01
Costs of variable inputs	0.38	0.19	0.34	0.06
Total annual costs	2.95	0.41	0.74	0.13

Source: Tables 3.3 and 3.4.
[a] Number of accessions in 1997.
[b] Projected number of accessions at capacity.

costly regeneration cycles. The high costs of durable inputs are the price paid for lower variable costs.

Second, the average variable costs comprise expenditures that would not be expected to rise per unit as accessions increase, for a given gene bank capacity. Thus the average cost curve lies above the marginal cost curve in the production range considered by this study. The average variable cost provides us with an upper-bound estimate of the marginal cost of conserving an accession. Conserving another accession in the gene bank costs only US$ 0.36 for wheat and US$ 5.83 for maize. In other words, if it costs more than US$ 5.83 to evaluate whether a maize accession is worth conserving, the evaluation is not economically justified; the decision should be to conserve the accession without further deliberation. Similarly, if it costs US$ 1.00 per accession to identify duplicate wheat accessions, such activity is not justified as a means of increasing gene bank efficiency.

The sharp difference between marginal and average total costs means that there are large economies to centralizing storage of all cultivars of a crop and avoiding excessive duplication of storage facilities. Given the relatively modest cost of "black box" or other forms of duplicate collections, security imperatives can be jointly satisfied with one central gene bank and several duplicates held in other parts of the world. At least one duplicate set should be at a location in which the prospects of political embargoes or military actions that could disrupt international access are extremely remote.

At CIMMYT, the distinct needs of wheat (and related small grains) and maize are met by one facility with provisions for both. Economies of scope appear to be significant, and advantages of learning about conservation by conserving both crops jointly also appear to be significant.

Finally, the large differences in average and marginal costs between wheat and maize accessions should be interpreted with care. The cost per accession is not the cost per unit of genetic diversity. Wheat accessions are typically highly homogenous, so that the diversity in a wheat gene bank is mainly between rather than within accessions. Open-pollinated maize, on the other hand, is highly heterogeneous. Each accession contains a wealth of genetic diversity. Indeed, the high cost of regeneration is related to the care that must be taken to maintain this diversity through several cycles of regeneration over coming centuries. Thus the cost per unit of diversity for maize is not necessary higher than for wheat. Comparisons along these lines are likely to be misleading.

Table 3.9. Summary of Variable and Durable Input Costs (US$), Total and per Accession

Crop	Total variable cost	Total fixed cost	Total cost	Number of accessions	Average variable cost	Average fixed cost	Average total cost
Wheat	44,537	179,574	224,111	123,000	0.36	1.46	1.82
Maize	99,151	160,398	259,549	17,000	5.83	9.44	15.27

Source: Calculations in Pardey *et al.* (1998).

3.5. RELATION TO OTHER GENE BANK COST STUDIES

The evaluation of the costs of the maize and wheat gene banks in the CIMMYT facility shows clearly that comparison of per-unit costs across different crop species is likely to be misleading as an indicator of the costs of conserving genetic diversity. Case studies for cassava (Epperson, Pachico, and Guevara, 1996) and sweet potato (Jarret and Florkowski, 1990) provide further information on the range of factors to consider in cost analyses.[5]

Cassava is traditionally conserved in a field gene bank, in which the crop is maintained by replanting year after year. The average total cost calculated by Epperson *et al.* was US$ 30.13, or about twice the cost of a maize accession calculated here, and about 25 times the cost of a wheat accession. But since the cassava field gene bank is highly labor intensive, the average variable cost difference is much greater; US$ 22.30 for the cassava field gene bank versus US$ 5.83 for maize and US$ 0.36 for wheat at CIMMYT. Since it is safe to assume that the marginal labor cost is non-decreasing, and the marginal costs at CIMMYT are non-increasing, the cost of adding an accession to the cassava field gene bank at the International Center for Tropical Agriculture (CIAT) is about 4 times the cost of adding a maize accession at CIMMYT and more than 60 times the cost for an added wheat accession. Although the methodologies employed in the two studies are not strictly comparable, the divergence between the cost estimates is so great that they indicate major differences in the underlying cost structure for conserving these two cereals and cassava.

Cassava is a very different crop from wheat or maize. Being a clone, the amount of diversity conserved per accession is also quite different. In addition, the technical setting of a field gene bank is distinct. The annual replanting of cassava can leave room for errors of identification that arise from confusion of adjacent plots, field pest attacks, and other factors.

A more instructive comparison is drawn by Epperson *et al.* between the cassava field gene bank just discussed and the *in vitro* cassava gene bank at CIAT. At CIAT, some of the facilities necessary for *in vitro* conservation are also used for virus cleaning, an operation required for worldwide distribution of germplasm from either gene bank. In addition, the specialized labor and structures required for the *in vitro* gene bank had excess capacity at current levels of operation and were therefore accounted as fixed costs. This means that although the average total cost per accession, at US$ 24.70, was of the same order as for the field gene bank (US$ 30.13), the average variable cost, at US$ 1.85, is less than 10% of the equivalent figure for the field gene bank (US$ 22.30). Assuming once more that average cost is an upper bound on marginal cost, the cost of placing an additional accession in each gene bank is dramatically different, though the average total costs are quite similar. Thus the benefit of using modern methods to detect duplication of accessions is dramatically higher if the accessions are duplicates in the field rather than *in vitro*.

Another earlier comparison of field and *in vitro* gene banks is the study by Jarret and Florkowski (1990) of the conservation of sweet potato (*Ipomoea batatas*) in Georgia. This paper contains a wealth of informative detail about the two conservation methods. The cost of the field gene bank is reported at US$ 28.00 per accession per year, exclusive of labor, and the equivalent cost of *in vitro* conservation is about US$ 22.00, which is similar to the average cost of the two CIAT gene banks discussed above. Unfortunately, the marginal costs, including labor and any necessary expansion facilities, are not explicitly reported.

3.6. CONCLUSION

In the modern *ex situ* gene bank at CIMMYT, the marginal cost of adding an accession is less than US$ 0.36/yr for wheat or US$ 5.83/yr for maize. In both crops, fixed costs are a major part of total costs. Centralization of storage of a given crop at one location can greatly reduce average costs of conservation, and economies of scope from storing different grains at a single facility also appear to be substantial.

Advances in technology have virtually eliminated the location-specificity of *ex situ* grain banks. Complete climate control enables independence from local weather, and advances in communications facilitate worldwide access to the bank through modern telecommunications and express mail facilities. If questions of security, freedom from phytosanitary controls, and political interference in access can be satisfactorily resolved, the argument for centralization of long-term conservation in only one site seems economically compelling. The physical security problem may be solved by present "black box" or other off-site arrangements for storing duplicates. If the political and phytosanitary risks are not eliminated, perhaps a second long-term world conservation facility can be justified.

The CIMMYT gene bank will not reach capacity in wheat or maize in the near future. Given this fact, the present value of the average cost of conserving another accession of wheat or maize *in perpetuity* to both the long-term and active collections is only US$ 0.36 / r = US$ 9.00 and US$ 5.83 / r = US$ 145.75 respectively, given an interest rate, r, of 4%. For wheat, this present value is almost certainly lower per accession than the cost of any competent economic evaluation of the value of maintaining an accession in the gene bank, were such an evaluation feasible and sufficiently accurate to be useful. The results for maize suggest that some decisions may have to be made in the future about the types of genetics resources to add to the collection. It may make more sense to conserve new accessions of landraces and wild relatives, for example, than recently created breeding lines.

Acknowledgments
This chapter is based on a paper commissioned by the CGIAR Systemwide Genetic Resources Program. Additional support was provided by the Swedish International Development Cooperation Agency. For assistance in collecting and interpreting data we are grateful to Arnoldo Amaya, Claudio Cafati, Jaime Díaz, Jesse Dubin, Tony Fischer, Paul Fox, Lucy Gilchrist, Arne Hede, Rafael Herrera, Alejandro López, Francisco Magallanes, Prabhu Pingali, and Eduardo de la Rosa. Vincent Smith provided valuable advice on aspects related to costing capital.

Notes

1 Senior authorship is not assigned.
2 Smith (1987) discusses various aspects related to the user cost of capital.
3 The choice of discount rate is crucial in these cost calculations. The 4% figure we chose is probably conservatively high, given the long-term return on other capital assets and the insurance-like function of the gene banks.
4 In the absence of meaningful information on any alternatives, we assumed away the prospects of technical change and other factors that could have changed the costs or service lives of these durable inputs.
5 Other cost studies include Godden *et al.* (1998) and Burstin *et al.* (1997).

References

Bailey, E. E., and A. F. Friedlaender. 1982. Market structure and multiproduct industries. *Journal of Economic Literature* 20: 1024–48.

Baumol, W. J., J. C. Panzar, and R. D. Willig. 1988. *Contestable Markets and the Theory of Industry Structure*. Orlando: Harcourt, Brace, and Jovanovich.

Burstin, J., M. Lefort, M. Mitteau, A. Sontot, and J. Guiard. 1997. Towards the assessment of the cost of gene banks management: conservation, regeneration and characterization. *Plant Varieties and Seeds* 10: 163–172.

Cooper, M.D., and N. A. deWath. 1976. The cost of on-line bibliographic searching. *Journal of Library Automation* 9: 195–209.

Crossa, J., S. Taba, S. A. Eberhart, P. Bretting, and R. Vencovsky. 1994. Practical considerations for maintaining germplasm in maize. *Theoretical and Applied Genetics* 89: 89–95.

Epperson, J. E., D. Pachico, and C. L. Guevara. 1997. A cost analysis of maintaining cassava plant genetic resources. *Crop Science* 37: 1641–9.

FAO (Food and Agricultural Organization of the United Nations). 1996. *Global Plan of Action for the Conservation and Sustainable Utilization of Plant Genetic Resources for Food and Agriculture*. Rome: FAO.

Falconer, D. S. 1981. *Introduction to Quantitative Genetics*. Second edition. Essex: Longman Group Limited.

Frankel, O. H., A. H. D. Brown, and J. J. Burdon. 1995. *The Conservation of Plant Biodiversity*. Cambridge, UK: Cambridge University Press.

Fuentes–Davila, G. 1996. Karnal bunt. In R. D. Wilcoxson and E. E. Saari, (eds.), *Bunt and Smut Diseases of Wheat: Concepts and Methods of Disease Management*. Mexico, D.F.: International Maize and Wheat Improvement Center (CIMMYT).

Godden, D., S. Wicks, J. Kennedy, and R. Kambuou. 1998. Decision support tools for crop plant germplasm maintenance in PNG (Papua New Guinea). Paper presented at the 42nd Annual Conference of the Australian Agricultural and Resource Economics Society, Armidale, New South Wales, Australia.

Hayami, Y., and V. W. Ruttan. 1985. *Agricultural Development: An International Perspective*. Revised and expanded edition. Baltimore: Johns Hopkins.

Jarret, R. L., and W. J. Florkowski. 1990. *In vitro* active vs. field gene bank maintenance of sweet potato germplasm: Major costs and considerations. *Horticultural Science* 25: 141–6.

Koo, B., and Wright, B. D. 1998. The effects of advances in biotechnology on the optimality of *ex ante* evaluation of gene bank material. Mimeo. Berkeley: Department of Agricultural and Resource Economics, University of California at Berkeley.

Leathers, H. D. 1991. Allocable fixed inputs as a cause of joint production: A cost function approach. *American Journal of Agricultural Economics* 73: 1083–1090.

Pardey, P. G., J. M. Alston, J. E. Christian, and S. Fan. 1996a. *Hidden Harvest: U.S. Benefits from International Research Aid*. Food Policy Report. Washington, DC: International Food Policy Research Institute (IFPRI).

Pardey, P. G., J. M. Alston, J. E. Christian, and S. Fan. 1996b. *Summary of a Productive Partnership: The Benefits from U.S. Participation in the CGIAR*. Environment and Production Technology Division (EPTD) Discussion Paper No. 18. Washington, DC: International Food Policy Research Institute (IFPRI).

Pardey, P.G., B. Skovmand, S. Taba, E. Van Dusen, and B. D. Wright. 1998. *Costing the* Ex situ *Conservation of Genetic Resources, Maize and Wheat at CIMMYT*. Draft Environment and Production Technology Division (EPTD) Discussion Paper. Washington, DC: International Food Policy Research Institute (IFPRI).

Smith, V.H. 1987. An econometric model of maintenance, utilization, scrapping and capital use in the U.S. electric power industry: Implications for the consequences of air quality regulation. Ph.D. thesis, North Carolina State University, Raleigh, North Carolina.

Wright, B. D. 1997. Crop genetic resource policy: the role of *ex situ* gene banks. *Australian Journal of Agricultural and Resource Economics* 41: 81–115.

APPENDIX
THE ANNUITY COST OF CAPITAL PURCHASED WITH REPLACEMENT

The present value of outlays on a capital item purchased and repurchased every n years is given by

$$PV_x = X + \frac{1}{(1+r)^n}X + \frac{1}{(1+r)^{2n}}X + ... + \frac{1}{(1+r)^{zn}}X$$

$$= X\left[1 + \frac{1}{(1+r)^n} + \frac{1}{(1+r)^{2n}} + ... + \frac{1}{(1+r)^{zn}}\right]$$

where zn is the last year of the purchase so that if $zn = \infty$ then $z = \infty$.

Letting $\dfrac{1}{(1+r)^n} = a$ then

$$\left[1 + \frac{1}{(1+r)^n} + \frac{1}{(1+r)^{2n}} + ... + \frac{1}{(1+r)^{zn}}\right] = \sum_{t=0}^{z} a^t$$

$$= \frac{(1-a)}{(1-a)}\sum_{t=0}^{z} a^t = \frac{1}{(1-a)}(1 + a^{(z+1)})$$

As $a < 1$, then $a^{(z+1)} \to 0$, as $z \to \infty$.

Thus as $z \to \infty,$ $\displaystyle\sum_{t=0}^{z} a^t = \frac{1}{(1-a)} = \left[\frac{1}{1-\left(\frac{1}{1+r}\right)^n}\right]$

so that $PV_x = \left[\dfrac{1}{1-\left(\dfrac{1}{1+r}\right)^n}\right] X$

Now consider the present value of an *annual* annuity Y over years 0 to zn.

$$PV_Y = \sum_{s=0}^{zn} \frac{1}{(1+r)^s} Y = Y \sum_{s=0}^{zn} \frac{1}{(1+r)^s} = Y \sum_{s=0}^{zn} \left(\frac{1}{1+r}\right)^s$$

Let $b = \left(\frac{1}{1+r}\right) < 1$

Then $\sum_{s=0}^{zn} = \left(\frac{1}{(1+r)}\right)^s = \frac{1-b}{1-b} \sum_{s=0}^{zn} b^s$

$$= \frac{1-b^{zn}}{1-b} = \frac{1}{1-b} \quad \text{if } z \to \infty$$

so that $PV_Y = \left[\dfrac{1}{1-\left(\dfrac{1}{1+r}\right)}\right] Y$

To calculate the annualized user cost, Y, of a capital item costing X repurchased each n years, set $PV_Y = PV_X$ and solve for Y in terms of X such that

$$\left[\frac{1}{1-\left(\dfrac{1}{1+r}\right)}\right] Y = \left[\frac{1}{1-\left(\dfrac{1}{1+r}\right)^n}\right] X$$

or

$$Y = \frac{\left[1-\left(\dfrac{1}{1+r}\right)\right]}{\left[1-\left(\dfrac{1}{1+r}\right)^n\right]} X$$

Thus with information on the real interest rate r, repurchase rate n years, and the cost of the capital item, X, the equivalent annualized user cost is given by Y. For example:

if $r = 0.04$; n = 40 years; $X =$ US\$ 1 million then $Y =$ US\$ 48,578, or

if $r = 0.04$; n = 15 years; $X =$ US\$ 1 million then $Y =$ US\$ 86,478.

4 OPTIMAL SEARCH IN *EX SITU* COLLECTIONS OF WHEAT

D. Gollin, M. Smale, and B. Skovmand

4.1. INTRODUCTION

The conservation of genetic resources for crop improvement has emerged as a controversial policy issue in recent years. In response to concerns about the perceived narrowing of the genetic base of crop plants (Shiva, 1993), the international community has spent millions of dollars to collect and conserve varieties of rice, wheat, maize, and other food crops. International institutions have devoted substantial resources to developing policies and protocols for the protection and exchange of crop genetic resources.

There are two principle strategies for preserving genetic resources for crop species, although the strategies are increasingly viewed as complements rather than as substitutes (Maxted, Ford-Lloyd, and Hawkes, 1997). One strategy is to preserve different varieties or species *in situ*. For cultivated species, *in situ* conservation as it is currently understood implies the management of traditional varieties by farmers in centers where the crop has evolved (see part III of this volume).

Another strategy is to preserve seeds or other propagative materials *ex situ*, in a collection that is physically separated from the environment of origin. Historically, *ex situ* collections of economically important crops date back hundreds of years. The early collections consisted of botanical gardens or simple fields in which different plant varieties were physically cultivated. Today, *ex situ* collections are more technologically sophisticated. For many crops, present-day gene banks are essentially huge refrigerators designed to store samples of seed at conditions of low temperature and low humidity for long periods. Such collections are principally intended to maintain genes for future crop improvements through plant breeding.

Ex situ storage remains the principal means of genetic resource conservation. Extensive *ex situ* collections already exist for most cultivated species. For wheat, the Food and Agriculture Organization (FAO) of the United Nations estimates that 95% of wheat landraces and 60% of wild species related to wheat have been collected. It is unclear on what data these figures are based, however, since the actual total number of populations or species is unknown. Similarly high figures are reported for other major crops, such as rice and maize, although wild species coverage is lower for these crops (FAO, 1996). Since the 1970s, many national and international research programs have expanded and upgraded their gene banks.

Wright (1997) has argued that the utilization of these genetic resources has not kept pace with their expanding numbers in gene banks. Critics have argued that *ex situ* collections are wasteful and poorly managed. One criticism, drawn from evolutionary biology, is that the materials stored in gene banks are "frozen" at the time of their collection, losing their potential to adapt to the changing natural and economic environments in which farmers produce crops (Guldager, 1975; see Frankel and Soulé, 1981). A second criticism, grounded in microeconomic principles, is that low utilization of accessions by breeding programs indicates that banks have excess capacity. Hence, marginal accessions are valueless. Most genetic resource specialists would agree that "accessions should be *used*, and breeders need to know what the packets or bottles of seeds on the shelves contain" (Plucknett *et al.* 1987: 74).

Do large collections ever have value? Are breeders and other scientists wastefully ignoring the materials in gene banks? Is breeders' low utilization of *ex situ* collections an indication of economic inefficiency? In this chapter, we begin to investigate such issues by using an economic and statistical framework based on search theory, and applying it to data on resistance to Russian wheat aphid (*Diuraphis noxia*) and septoria tritici leaf blotch.[1] Frequency distributions obtained from the Genetic Resources Information Network[2] and CIMMYT scientists are used to estimate "smooth" representations of the underlying distributions for resistance. We combine this information with estimates of the benefits that can be attained from finding resistant materials and with information on the cost of searching for resistant materials. The model then guides us in identifying optimal strategies for searching the gene bank and for utilizing different categories of materials. We use Monte Carlo simulations to solve the model computationally.

Three specific questions are posed and answered in the analysis. We begin by using the model to determine how many gene bank accessions should optimally be included in a search for a useful trait, given a category of genetic resources. We then estimate the value of specialized knowledge about the distribution of desirable traits across types of germplasm. In this case, we compute the value of knowing that resistance to Russian wheat aphid is more common among a set of bread wheat landraces from Iran than among the general population of bread wheats. Finally, we ask how an optimal search should proceed when scientists can hunt for a particular trait among two distinct populations with differing distributions of resistance, using the case of septoria tritici leaf blotch as an illustration. While each application of the model represents a special case, the findings provide some general conclusions regarding the utilization of gene banks and the valuation of accessions.

The next section of this chapter briefly reviews relevant research. This review is followed by a brief presentation of the economic decisionmaking rule implied by the search theoretic model. Next, we describe the data, the solution approach, and results of the experiments. Conclusions and their implications for further research are discussed in the concluding sections.

4.2. RELATED SEARCH MODELS

There is little confusion or disagreement over the concepts to apply in economic analysis of genetic resources; marginal analysis suffices (Brown, 1990; Evenson, 1993; Pearce and Moran, 1994; Swanson, 1995). Searches of a collection should proceed until the marginal expected value of success is outweighed by the search cost.

Evenson and Kislev (1975) used a search model to analyze the economics of "discovering" new crop varieties in agricultural research. Using data on sugarcane research, they answered questions about optimal search strategies, investigated the effects of changes in research technology, and explored policy issues for research. An implication of their model was that additional investments in research could hasten the rate of "discovery" of improved varieties, reduce search costs, or both.

More recently, Simpson, Sedjo, and Reid (1996) applied a search model to the problem of valuing a marginal species in a tropical rainforest. In this model, pharmaceutical researchers test a large number of species for a particular trait of economic value. The trait is assumed to be distributed randomly and uniformly across the entire population: with a given probability, each species either possesses the trait or fails to possess the trait. The search process thus consists of repeated and independent Bernoulli trials. Using data from the pharmaceutical industry, the authors concluded that under most plausible specifications, the expected value of a marginal species is minuscule. The authors observed that "*regardless* of the probability with which the discovery of a commercially useful compound may be made, if the set of organisms that may be sampled is very large, the value of the marginal species must be very small" (emphasis in the original). Simpson and Sedjo (1998) developed a similar model for genetic resource conservation in agriculture, although they did not apply the model empirically.

Simpson, Sedjo, and Reid thus expressed a relatively pessimistic view towards the marginal value of genetic resources, in contrast to the generally optimistic implications of the Evenson and Kislev model. The differences between their conclusions indicates the sensitivity of search theoretic models to important assumptions concerning the *redundancy* of materials in a collection of objects to be searched.

Suppose that a desirable trait is found with some distribution within a particular population, and suppose that it is equally useful wherever it is found. For example, suppose we are seeking flowers that are yellow, and that any shade of yellow is equally useful. In this case, there is perfect substitutability among the subpopulation of species that possesses the trait (all yellow flowers), and zero substitutability with the remainder of the population. In such a case, once a single yellow flower is found, all further search becomes redundant. This is the key assumption in Simpson, Sedjo, and Reid (1996).

Alternatively, suppose that the desired trait can be found in varying intensities or forms, so that it can be conceptualized as a continuous variable. For example, gold might be found in different deposits of ore at varying degrees of concentration, or a number of different plants might have antibacterial properties of varying usefulness. In this case, there is imperfect substitutability among materials in the population. Different materials are more or less desirable, according to some index, and there is some distribution of "desirability" across the population. In this case, additional searching is always expected to offer some marginal benefit unless some extreme value has been obtained. This is the result obtained by Evenson and Kislev (1975).

To answer questions about searching *ex situ* collections of crop genetic resources, we need a model that does not arbitrarily impose any assumption about the nature or the distribution of desirable traits. The approach used here is based on assumptions similar to those of Evenson and Kislev. The underlying model allows for continuous distributions for the genetic expression of a trait and implies imperfect substitutability among bank accessions. To represent the specific case of screening for disease resistance in wheat, however, we identify a level of "useful" resistance from a distribution of scores. We define the probability of "success" as the probability of finding an accession with a "useful" level of resistance, and the draws of accessions resemble repeated Bernoulli trials.

4.3. OPTIMAL SEARCH WITH PERFECT INFORMATION

What is the optimal size of a search for resistance when the probability distribution of resistance is "known" by the researcher? A researcher who knows the distribution of resistance would choose the size of the search such that the expected marginal benefit of search equals the marginal search cost. Let the function $H_{n_j}^{j}(s^*)$ give the probability that a search of size n among materials of type j will result in a "useful" discovery, where the usefulness of resistance is defined by having a score below s^*. Material j is a particular subpopulation of materials from the set J of possible subpopulations, such as a subpopulation of landraces, obsolete cultivars, wild relatives, or breeding lines. We also know that H depends on the parameters of the underlying distribution of resistance, $\Delta^j(S)$. Denote these parameters by θ. To emphasize the function's dependence on n, we can write the function as $H_{n_j}^{j}(n;s^*,\theta)$.

Then the optimal size of search within a subpopulation j is given by:

$$(4.1) \quad \left[\sum_{m \in M} \sum_{t=T_j+\bar{T}_m}^{T_j+\bar{T}_m+\tau} \beta^t \delta(t) \left(Ev_m(z_j) - v_m(\bar{z}) \right) \right] \times \frac{\partial H^j(n;s^*,\theta)}{\partial n} = c.$$

The left-hand side in this equation represents the marginal benefit of expanding the search, with the first term giving the total benefit stream and the second term giving the marginal change in the probability of successful search when n is increased. Note that under our assumptions regarding the search for resistant materials, $H_{n_j}^{j}(n;s^*,\theta)$ can be reduced to the Bernoulli expression $1-(1-p)^n$. Resistance is treated here as a "threshold" characteristic, rather than as a continuous variable.

The total benefit stream represents the sum of the discounted annual savings over the average value of production given the incorporation of the trait $[v_m (\hat{z}) - v_m (\bar{z})]$ across the set of M "mega-environments." CIMMYT, for example, classifies the world bread wheat growing area into "mega-environments" that span different continents and regions within the developing world. These environments are distinguished from one another based on the kinds of wheat technologies they require. As new biotypes of pathogens emerge over time, the productivity advantage of a novel source of resistance follows a depreciation path $\delta(t)$. The discount factor β corresponds to the rate of time preference of the decision-makers represented in the model.

The timing of the search process, the release of a new, resistant variety, and that variety's diffusion all have implications for the benefits stream. A time lag is associated with transferring resistance genes from materials identified in the evaluation process into a breeding program, and this lag varies depending on the type of material and the heritability of the trait (T_j). This period is often called "pre-breeding." A second lag extends from the time a trait is transferred to the breeding program to the time at which diffusion of a new variety with the trait begins in farmers' fields (T_m). This time lag differs by mega-environment. Varieties for areas that favor wheat production may be developed more quickly than varieties destined for marginal wheat production environments. Further, because of other factors related to farmers' adoption of new wheat varieties, the diffusion of materials that carry the resistance may begin later in some areas than in others. Finally, let τ denote the time horizon of relevance to breeders.

The right-hand side of the equation represents the marginal cost of search, c. For the choice of the optimal number of accessions to search within a category of material j, the relevant costs are those associated with screening and evaluating an additional accession. For the questions posed here, most of the costs of operating a gene bank—including the costs of construction, power, and day-to-day management—can be viewed as fixed, in the sense that they do not vary with the size of a search or with the type of materials searched. The costs of the breeding program are also fixed from the standpoint of the search decision.

4.4. SOLUTION APPROACH AND DATA

In the experiments described here, we equated the marginal benefits to marginal costs of search, using (1) the probability distributions for "useful" resistance generated with Monte Carlo simulations from smoothed, actual distributions for the trait, (2) estimates of benefit streams, and (3) representative cost data. Methods used to develop each of these are summarized below and described in greater detail in Gollin, Smale, and Skovmand (1998).

To apply the model, we need data on the actual distributions of traits within the relevant population and on the costs and benefits of searching for such traits. Recent studies in applied genetics focus on molecular differences among varieties and compare small segments of DNA (Autrique *et al.*, 1996; Tsegaye, Becker, and Tesemma, 1994; Pecetti and Damania, 1996; Chen *et al.*, 1994). Molecular analysis of genetic diversity in wheat has not yet advanced to the point where it is informative about the distributions of rare alleles within large populations of wheat varieties. While Tanksley and McCouch (1997) have argued that phenotypic evaluation to determine the breeding value of an accession is likely to be misleading (especially with respect to quantitative characteristics), it is the best tool available at present. Here, we use data measured on phenotypic expression.

4.4.1. Probability Distributions for Traits

The population of wheat accessions found in national and international collections can be thought of as a large draw from the distribution of wheat populations found worldwide. The subpopulations evaluated for resistance to particular disease or pest problems are then second-stage samples from the gene banks. Given an actual discrete distribution of scores for disease or pest resistance over any subcollection of wheat varieties, we can calculate the distribution $H_{n_j}^j$ $(n;s^*,\theta)$.

The Genetic Resources Information Network provides data on discrete distributions for certain traits within types of wheat accesions, taken from performance scores in various agronomic trials, represented in scales. These scales are essentially imposed for convenience by agricultural scientists on an underlying, "true" distribution which we have reconstructed. A least squares technique was used to fit a smooth, beta distribution to each discrete distribution of disease and pest resistance scores. The parameters of a beta distribution are generally written as a, b, and β. Two of the three parameters are independently determined, and the third can be computed from the other two. The beta distributions can take on a wide range of shapes and forms, and can be used with data that range from approximately symmetric to severely skewed. Beta distributions can also allow for tails of varying thickness, which is important for the kinds of analysis undertaken here. The beta distributions offer good representations of the data, although the goodness of fit varies.

4.4.2. Benefits

We estimated the areas affected by diseases, average annual yield losses, and expected lifetime of resistance based on published literature and personal communication (CIMMYT, 1985; Dubin and Rajaram, 1996; Great Plains Agricultural Council on Russian Wheat Aphid, 1989; Marasas *et al.*, 1997; Rajaram and van Ginkel, 1996; Scharen and Sanderson, 1985; Robinson, 1994; Gilchrist and Mujeeb-Kazi, 1996; J. Dubin, personal communication; L. Gilchrist, personal communication). Average yields by environment were obtained from the Wheat Impacts databases of the CIMMYT Economics Program and were adjusted for yield potential according to estimates provided in Sayre *et al.* (1998) and Byerlee and Moya (1993).

"Crop losses averted" were used as an estimate of the benefits associated with achieving an improved level of resistance. Losses averted in each time period have been calculated as:

$$(4.2) \quad q_t = \sum_m a_{mt} \gamma_{mt} y_{mt}$$

where a_{mt} is the area planted to varieties carrying resistance in environment m in time t, γ_{mt} is the average annual percent yield loss from disease over the area usually affected by disease in that year, and y_{mt} is the average wheat yield in farmers' fields.

The yield losses averted decrease over the time period t as the source of resistance depreciates or decays with the evolution of new disease pathotypes, as defined by $\delta(t)$. The areas planted to varieties carrying resistance follow diffusion paths that differ by environment. Adoption ceilings are based on the proportion of materials grown in that environment in 1990 that were CIMMYT crosses or had at least one CIMMYT parent. The period over which the benefits are calculated depends on the type of material from which the source of resistance was obtained, and the environment in which the resistant varieties are grown. The average research lag in each mega-environment covers the time from entry of the resistant lines into the breeding program, through the development of finished varieties, until the time that farmers begin to grow them (T_m).

Our estimates are based on a number of assumptions, most of which result in an understatement of the magnitude of benefits. First, we computed the benefits of finding useful materials on the assumption that CIMMYT performs the search, incorporates the materials into breeding lines, and disseminates the resulting breeding lines to national programs. The benefit streams are based only on the major environments in which spring bread wheats are grown in the developing world. Our calculation ignores: (1) the potential for incorporating resistance into other wheat types; (2) other avenues through which useful materials might pass into national programs and farmers' fields; (3) benefits obtained through the subsequent diffusion of varieties that incorporate the resistance trait through more distant CIMMYT ancestors; and (4) benefits transferred through trade to consumers in countries not included in CIMMYT's mandate.

Second, we consider only a 35 year time horizon, since depreciation of resistance and discounting imply that benefits obtained beyond this time horizon are of little immediate consequence. It is possible to extend the analysis to incorporate longer time periods, but there are few data on which to base this analysis, and quantitatively speaking, benefits occurring outside this time period are of little consequence, given discount rates of the magnitude most widely used.

Third, we assume that the diffusion of varieties carrying a new source of resistance would follow a trajectory similar to the diffusion pattern of semidwarf wheats that was observed between 1967 and 1990. Chances are, however, that today's diffusion paths would have a steeper slope as a result of improvements in seed systems and increased commercialization of wheat growers in many environments. In our simulations, we assume the same differences in diffusion between marginal and favorable environments that occurred historically.

As compared to a more comprehensive social impact analysis that considers the effects of technical change on consumers, related factor or output markets, and health or environmental amenities, our analysis considers only production benefits. Note that genetic resistance to disease would also generate benefits, however, by making it possible to avoid the negative health and environmental effects of chemical disease control.

Finally, we assume that the evolution of new disease and pest biotypes can be captured adequately through the simple structure used here to model the breakdown of resistance over time as a constant rate of depreciation. The diffusion of resistant varieties can actually contribute to the emergence of new disease and pest pathotypes, which we treat as an exogenous event. The simplicity of our approach reflects the fact that its objective is to analyze the economics of search rather than to assess research impacts.

4.4.3. Costs

Cost estimates and the conceptual basis of the cost structure in this study are based on historical experience with the CIMMYT wheat gene bank. Search costs were recorded in terms of three components: general program costs, evaluation costs, and the costs of transferring useful genes from the varieties in which they are initially found into breeding lines. Program costs, such as salaries and equipment, are roughly proportional to the size of the search undertaken. Evaluation costs include expenses for land and crop management, preparation of insects or inoculum, and note-taking. These costs are also linear in the size of search. They generally do not differ with the types of materials being evaluated, but they may vary according to the characteristics of the disease or pest.

By contrast, transfer costs—also known as "pre-breeding costs"—vary with the different types of materials used and the nature of the trait. They usually decrease with an increasing level of improvement in the source material. Simple traits and those that are highly heritable are relatively easy to transfer, while complex traits and traits that are not highly heritable are difficult to transfer. At one extreme, for example, no pre-breeding is required to transfer a trait that is found within a released variety or an elite line. At the other extreme, a wild relative of wheat has high transfer costs because special techniques are required to make it useful to the crossing program. Applied molecular biological techniques are expected to reduce the costs of pre-breeding for such materials in the near future, but there will always be a range of pre-breeding costs for different sources of genetic resistance and traits.

Transfer or pre-breeding costs are best thought of as fixed costs incurred on a one-time basis when material is transferred into breeding lines. Once the desirable trait has been moved into an advanced line, it may be used freely. Since transfer costs vary by germplasm type rather than the number of accessions screened, they are an important consideration in the choice of material to search, rather than in the size of the search.

A second type of fixed cost of search might include the costs of developing search techniques or setting up a new experiment. The fixed costs of search might differ across various types of material if, for example, different techniques are used to screen for resistance among wild species.

In the work that follows, variable costs are assumed to fall in a lump sum in the first year of the time period ($t=0$) rather than at T_j, leading us to overestimate their magnitude slightly. Transfer costs are treated as fixed costs incurred at time zero, but in fact neither transfer costs nor the fixed cost of search were quantitatively significant in our experiments. The *time* to transfer and the *time* in the breeding program are critical in determining the discounted net benefits stream, however. These clearly vary by type of material, transfer technique, and breeding program.

4.5. RESULTS

4.5.1. An Illustration of Optimal Search
We use the example of Russian wheat aphid (RWA) to illustrate the solution to the search problem. From its center of origin in the Caucasus and Central Asia, the aphid has emerged as a pest of some importance in the United States, the Republic of South Africa, parts of the Southern Cone of Latin America, and North and East Africa. The pest is potentially important in Australia and parts of the People's Republic of China (Robinson, 1994). Some (but not all) of the wheat lines from the countries of origin of the pest have resistance (Marasas *et al.*, 1997).

Searches among bread wheat varieties in the United States Department of Agriculture (USDA) collection yielded almost no useful material, although a few wheats displayed effective resistance. Of 41,109 wheat accessions evaluated by the USDA—most of them elite lines and released varieties—just over 100 displayed any useful resistance. None was a spring-habit bread wheat. Literature summarizing searches for resistance also reports the near absence of sources of resistance in improved materials or any materials originating outside of Central Asia (Robinson and Skovmand, 1992; Souza *et al.*, 1991; Harvey and Martin, 1990; du Toit, 1987). The question facing CIMMYT researchers was how best to search their collection to find little-known sources of resistance.

Figure 4.1 shows both the raw histogram and the smoothed beta distribution that approximates the underlying distribution of resistance among 10,190 landraces for which data were available in the Genetic Resources Information Network. The distribution has a very thin left-hand tail; almost no resistant landraces were found.

Table 4.1 shows total discounted net benefits for the incorporation of resistance to RWA into CIMMYT bread wheat materials under several sets of assumptions concerning: (1) areas affected by the pest; (2) average yield losses; (3) longevity of resistance; and (4) the time lags associated with pre-breeding and breeding. Benefits are modest in CIMMYT's major mega-environments. In CIMMYT's mandate area,

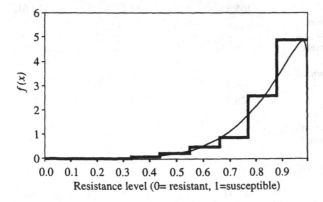

Figure 4.1. Actual Distribution of Resistance to Russian Wheat Aphid in Landrace Accessions, with Smooth Fitted Approximation.

Note: Actual distribution based on Genetic Resources Information Network data for 10,190 landraces of *Triticum aestivum*. Smooth approximation is a beta distribution fitted by least squares.

yield losses are minor because RWA is a major problem only in small areas that occur in marginal production environments. Benefits are much larger in magnitude when based on the full global area delineated by the CLIMEX model (reported in Robinson, 1994), although our estimation methods remain conservative. For example, RWA reportedly caused a cumulative loss of US$ 890 million in the United States alone from 1987 to 1993. Only 39% of the loss was attributed to lost production. About 9% was spent on chemical control, and 52% was estimated as lost in other economic activities. Negative externalities of chemical use were not considered in that analysis. In their comprehensive analysis of the impact of the Russian Wheat Aphid Control Research Program in the Republic of South Africa, Marasas *et al.*(1997) estimated a 34.6% internal rate of return to the yield savings associated with the investment in developing resistant cultivars, for the period 1980–2005. The estimates presented here are based solely on yield savings.

A key analytical result demonstrated in Table 4.1 is that the time lag for transfer, breeding, and adoption affects the economic return by a large order of magnitude. Transferring a source of resistance from a landrace using conventional breeding techniques would undoubtedly require considerably more than one or two years. The breeding lag usually assumed for CIMMYT is five years, while ten years may be a more reasonable assumption when materials are developed by CIMMYT and then used or adapted by national breeding programs.

In all scenarios, the cost of search is estimated at US$ 82.97 per landrace screened (Skovmand, unpublished data). Based on the experience that variable search costs do

Table 4.1. Benefit Streams Associated with Finding a Source of Resistance to Russian Wheat Aphid in Bread Wheat Landraces (US$ million 1990)

Assumptions and parameters	Transfer lag: 2 yr Breeding lag: 5 yr	Transfer lag: 5 yr Breeding lag: 10 yr	Transfer lag: 10 yr Breeding lag: 10 yr
Global			
23 m ha affected	165.82	48.97	18.61
8.3 m ha ceiling adoption			
5% initial average yield loss			
15 yr duration of resistance			
Variable adoption lag			
Major CIMMYT spring wheat			
mega-environments			
9 m ha affected	29.2	9.97	3.77
3.4 m ha ceiling adoption			
5% initial average yield loss			
15 yr duration of resistance			
No adoption lag			
Major CIMMYT spring wheat			
mega-environments			
4 m ha affected	15.67	3.315	1.2
1.6 m ha ceiling adoption			
5% initial average yield loss			
15 yr duration of resistance			
No adoption lag			

not change substantially with the size of the search, we have assumed that all costs are variable costs, resulting in a constant average search cost equivalent to a constant marginal cost.

From these data, it is possible to compute the optimal search size for alternative scenarios. Figure 4.2 shows the relationship between search size and the probabilities of successful search for resistance to RWA among accessions of bread wheat landraces. Figure 4.3 combines this information with the associated marginal benefits and marginal costs. When the least favorable assumptions about the benefit stream are used, the optimal search size is about 4,700 landrace accessions, illustrated by the point at which the marginal benefit curve intersects the marginal cost curve in Figure 4.3a. The most conservative scenario assumes initial average yield losses of 5% on a total of 4 million affected hectares, with an adoption ceiling of only 1.6 million hectares, and a 15 year longevity of resistance. The time lag for transferring resistance is ten years, with an additional ten year research lag for breeding (Table 4.1). In this scenario, the expected total benefits of finding a landrace with resistance to RWA are US$ 865,000 with a total cost of US$ 406,000, for total net benefits of US$ 459,000.

The intermediate scenario presented in Figure 4.3b also assumes a long transfer and breeding lag, a 5% annual average yield loss, and resistance lasting 15 years, but the adoption ceiling in this scenario is 3.4 million hectares. The optimal size of search is about 10,000 landraces accessions, with an expected total benefit of US$ 3,452,000, and total cost of US$ 830,000, for a total net benefit of US$ 2,622,000.

The benefit streams in both the least favorable and intermediate scenario are based on CIMMYT spring wheat mandate areas alone, and the longest research lag. Under the global benefit scenario with the longest research lag, the benefits are large enough to justify a search of approximately 18,000 landraces (Figure 4.3c). Total search costs are US$ 1,493,000 and expected benefits are US$ 18,228,000, for an expected net benefit of over US$ 16,735,000.

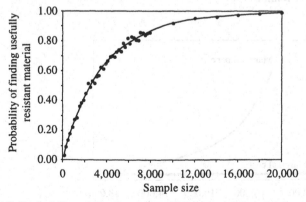

Figure 4.2. Probability of Finding Usefully Resistant Material from Draws of Given Sample Size, Monte Carlo Simulations and Smoothed Function.
Note: Data from Monte Carlo simulations are based on repeated draws from the actual distribution of
   resistance to Russian wheat aphid among landrace accessions, with smoothed function. The
   smoothed function is a sixth-order polynomial, fitted by least squares.

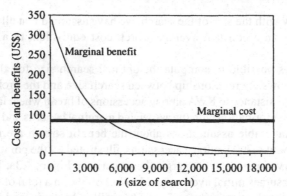

Figure 4.3a. Marginal Costs and Marginal Benefits of Searching for Russian Wheat Aphid Resistance in a Sample of *Triticum aestivum* Landraces, Low Scenario (total benefits = US$ 1.2 million).

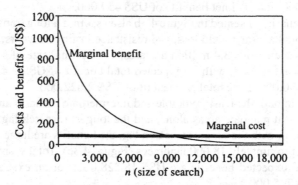

Figure 4.3b. Marginal Costs and Marginal Benefits of Searching for Russian Wheat Aphid Resistance in a Sample of *Triticum aestivum* Landraces, Intermediate Scenario (total benefits = US$ 3.77 million).

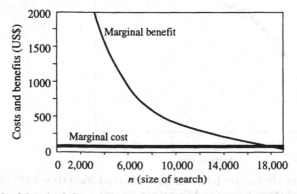

Figure 4.3c. Marginal Costs and Marginal Benefits of Searching for Russian Wheat Aphid Resistance in a Sample of *Triticum aestivum* Landraces, High Scenario (total benefits = US$ 18.23 million).

When global benefit streams are assumed with a total transfer and breeding lag of only seven years (Table 4.1), the economic problem becomes trivial: the optimal size of search is larger than the number of landraces in the CIMMYT gene bank. The benefits are so great relative to costs that a search of all existing accessions would be justified. This result does not tell us whether it would be worthwhile to collect additional accessions, however, since we have not considered the costs of collection and storage.

Given any total benefit stream, we can compute the optimal search size n^* and the associated expected net benefits. As Table 4.2 indicates, optimal search size increases with the total benefit stream.

Table 4.2. Optimal Search Size and Expected Net Benefits as Determined by the Size of Total Benefit Stream, for the Case of Russian Wheat Aphid

Total benefits (US$)	Approximate optimal search size (n^*)	Expected net benefits (US$)
500,000	1,750	46,388
1,000,000	4,100	321,428
5,000,000	10,500	3,752,204
10,000,000	12,650	8,456,173
15,000,000	14,150	13,236,039
20,000,000	18,900	18,153,771
25,000,000	19,150	23,086,572
30,000,000	14,000	28,022,376

4.5.2. The Value of Specialized Knowledge

Our next experiment asks: What is the value of specialized knowledge about the distribution of desirable traits across various subpopulations of wheat varieties? Figure 4.4 displays the actual distribution of resistance to RWA in a subset of 1,089 Iranian landraces evaluated by the CIMMYT germplasm bank. This distribution implies that a very small search is likely to yield resistant materials; as shown in Figure 4.5, a search of 15 accessions was almost certain to result in usefully resistant materials. For the

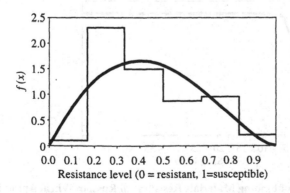

Figure 4.4. Beta Distribution to Approximate Actual Pattern of Resistance to Russian Wheat Aphid in 1,089 Iranian Landraces of *Triticum aestivum,* as Fitted by Least Squares Approximation Technique.

most favorable scenario, expected net benefits rise by US$ 1,690,000 when Iranian landraces are searched instead of the entire landrace population. For the intermediate scenario, the expected net benefits rise by US$ 1,110,000 with specialized knowledge; for the lowest scenario, the expected net benefits rise by US$ 73,000. At least two benefits are obtained if a gene bank manager knows how to focus a search: savings on search costs and a greater probability of finding useful material. Both of these contribute to the increase in expected net benefits. The time lags of research discovery may also be affected.

Although it is clear that specialized knowledge can be extraordinarily valuable, we do not make any claim as to its "uniqueness." It may be that many people share the specialized knowledge, so that we are measuring the value of publicly available information. In many cases, though, it seems reasonable to imagine that the specialized knowledge is held by a relatively small number of scientists.

4.5.3. Searching for Resistance in Multiple Categories of Material
In most cases, researchers have the option of searching for desirable traits in more than one category of germplasm, such as landraces, elite lines, and wild relatives. This final experiment asks: What is the best way to allocate search resources among types of germplasm?

If more than one category of material is searched, or an interior solution holds, then the efficiency conditions of economic theory require that the expected marginal benefits of search should be equalized across categories of germplasm. Commonly, however, such problems have corner solutions: when the distribution in each material is known, optimal search will omit all but the category with the highest expected marginal benefits. The researcher's problem is to select the material that gives the highest overall returns.

To analyze this question, we consider the case of resistance to *Septoria tritici,* a pathogen causing a leaf blotch that affects wheat on over 10 million hectares worldwide. Most of the affected area is found in the CIMMYT mandate areas (9 million hectares), including large portions of the Southern Cone of South America (Brazil, Chile,

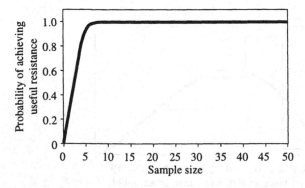

Figure 4.5. Probability of Finding Materials Resistant to Russian Wheat Aphid in Draws from a Population of Iranian Landraces.
Note: Based on Monte Carlo simulations using a smoothed version of the actual distribution of resistance in a population of 1,089 Iranian landraces screened by the CIMMYT wheat gene bank.

Argentina, and Uruguay). In a research initiative to diversify the genetic basis of resistance to septoria tritici leaf blotch, CIMMYT scientists searched for new sources of resistance in breeding lines, landraces, and other materials, including emmer wheat (*Triticum dicoccon*) (Fuentes and Gilchrist, 1994; Gilchrist and Mujeeb-Kazi, 1996; Gilchrist and Skovmand, 1995). Here, we consider the comparison of breeding lines and emmer wheat only. We use data on resistance from searches conducted by the CIMMYT germplasm bank. As above, we assume that researchers have full information about the distributions of resistance within these two populations.

To ask how a search would proceed optimally with two types of materials, we need to know the distributions of resistance, benefit streams, and search costs associated with each type. The distributions of resistance to septoria tritici leaf blotch are shown in Figures 4.6 and 4.7. Almost all of the emmer accessions that were tested displayed useful resistance, whereas relatively few of the breeding lines did so. The distribution of resistance in emmer wheat thus dominates the distribution of resistance in breeding

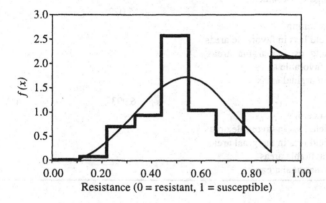

Figure 4.6. Actual and Fitted Distributions of Resistance for Septoria tritici Leaf Blotch in 1,834 Breeding Lines.

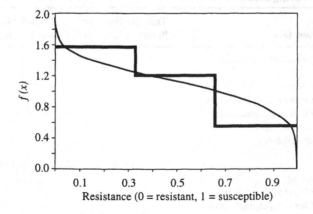

Figure 4.7. Actual Distribution of Resistance to Septoria tritici Leaf Blotch in 1,729 Accessions of Emmer, with Smooth Fitted Approximation.

lines, in the first-order stochastic sense. This might suggest that scientists should ignore the breeding lines altogether. But there are important differences in benefit streams between the two types of material (Tables 4.3 and 4.4), resulting essentially from differences in the time lags associated with transferring the resistance into advanced breeding materials. The higher benefit stream is associated with finding resistance in breeding materials. Then the transfer of resistance into other breeding materials is only two years, by a conservative estimate, as compared to the five years needed to transfer resistance from emmer wheat to breeding lines.

Table 4.3. Benefit Streams Associated with Finding a Source of Resistance to Septoria tritici Leaf Blotch in Breeding Materials of Spring Bread Wheat (US$ million 1990)

Assumptions and parameters	Transfer lag: 2 yr Breeding lag: 5 yr	Transfer lag: 2 yr Breeding lag: 10 yr
Major CIMMYT mega-environments		
9 m ha affected	15.022	7.989
5.5 m ha adoption ceiling		
2% initial avg. yield loss in favorable areas		
1% initial avg. yield loss in marginal areas		
10 yr duration in favorable areas		
15 yr duration in marginal areas		
7 m ha affected	6.993	4.25
4.1 m ha adoption ceiling		
2% initial avg. yield loss in favorable areas		
1% initial avg. yield loss in marginal areas		
7 yr duration in favorable areas		
10 yr duration in marginal areas		

Table 4.4. Benefit Streams Associated with Finding a Source of Resistance to Septoria tritici Leaf Blotch in Emmer Wheat (US$ million 1990, assuming resistance is transferred only into spring bread wheat)

Assumptions and parameters	Transfer lag: 5 yr Breeding lag: 5 yr	Transfer lag: 5 yr Breeding lag: 10 yr
Major CIMMYT mega-environments		
9 m ha affected	11.854	4.358
5.5 m ha adoption ceiling		
2% initial avg. yield loss in favorable areas		
1% initial avg. yield loss in marginal areas		
10 yr duration in favorable areas		
15 yr duration in marginal areas		
7 m ha affected	6.376	2.316
4.1 m ha adoption ceiling		
2% initial avg. yield loss in favorable areas		
1% initial avg. yield loss in marginal areas		
7 yr duration in favorable areas		
10 yr duration in marginal areas		

The variable costs of searching for resistant breeding materials are also much lower than the comparable costs of searching among emmer materials. It is relatively quick and easy to evaluate breeding materials for resistance: they can simply be subjected to disease stress, and resistant materials can be selected. Accessions of emmer wheat must first be head-selected to remove heterogeneity, and it is more difficult and time-consuming to grow the plants and subject them to the necessary stresses because they are taller and less uniform. For the case of septoria tritici leaf blotch, average variable search costs have been estimated at about US$ 6 per accession for breeding material and about US$ 80 per accession for emmers.

The optimal search strategy in this case is surprising. Despite the favorable distribution of resistance among accessions of emmer, it is optimal to search only within the category of breeding materials; i.e., to select a corner solution. To see this, suppose that we could be certain of finding an emmer accession with useful resistance in a draw of only one accession. Using the most favorable set of estimates, this would yield benefits of US$ 6,376,000 (less US$ 80 in search costs). But within the collection of breeding materials, a search of at most 220 varieties will have a 99% probability of identifying usefully resistant materials.[3] The costs of this search are trivial (at most US$ 912), and the benefit stream is higher, because the yield gains are attained sooner due to the ease of transferring the resistance into new varieties. The net benefit is US$ 6,992,000, exceeding the expected benefits from the emmer wheats. Under this scenario, then, it does not make sense to search for resistance among the emmer accessions if the same trait can be found in the breeding lines. Other benefit scenarios give rise to the same conclusion.

Although the distribution of *S. tritici* resistance is more favorable in emmer accessions than in breeding materials, the difference in benefit streams is of overriding importance. From an economic point of view, the breeding materials strictly dominate the accessions of emmer wheat for resistance to *S. tritici*.

In other cases, however, it may make economic sense to search in multiple categories of genetic resources, even if the payoffs are similar to those described for septoria tritici leaf blotch. For example, it may be cheaper to search a second category of materials, given that a search is already being conducted. A search can help achieve more than one objective, such as a scientific advance, the testing of a new technique, or the screening of the same material for several traits simultaneously.

It will not always be true that researchers are better off working with breeding materials. The distributions of some kinds of resistance are superior in landraces or wild species. When a desired trait is absent among breeding materials, or when the resistance distributions are distinctly better among unimproved materials, it make senses for researchers to focus their search on landraces. Given the search cost differentials and time lags, however, it is frequently rational for researchers *not* to use unimproved materials from the gene bank. This finding explains why plant breeders, pathologists, and other scientists sometimes appear reluctant to use such materials.

In some cases, of course, researchers will turn to unimproved materials in an effort to "broaden the base of resistance" to a particular disease or pest problem—in other words, to find alternative resistance genes that would substitute for a source of resistance they have already identified. This does not violate the principle described above: to seek a new source of resistance is essentially to search for a new trait. The expectation is that the new trait will be distributed more favorably among unimproved materials than among breeding lines. In such cases, it makes sense to search among unimproved materials despite the longer time lags and higher costs. This was in fact the nature of the search at CIMMYT for resistance to septoria tritici leaf blotch: the goal was to find new sources of resistance that might allow breeders to diversify its genetic base.

4.6. CONCLUSIONS

The experiments described in this chapter result in three conclusions related to the management and valuation of *ex situ* collections of genetic resources. One conclusion is that the optimal scale of a search for desirable traits is very sensitive to the size of the economic problem as well as to the probability distribution for the trait. For some traits, the payoffs are simply not large enough to justify exhaustive searches. For other traits, the distributions are such that small searches will suffice. There are occasional situations, however, in which the distribution of resistance and the payoffs to discovery are such that large searches are justified. These are the situations when large collections are valuable, as in the case of resistance to RWA.

A second conclusion is that differences across types of genetic materials in the cost of search and in the associated time lags can lead to optimal search strategies in which some materials are systematically ignored. For example, unless the probability of success with emmer wheats is dramatically higher than with breeding lines, as long as the time lags associated with incorporating desirable traits from them are higher than the time lags associated with breeding lines, accessions of emmer wheat will be excluded. Given conventional breeding techniques, the fact that many gene banks have been little used is economically rational—and it says nothing about their long-term value as technologies as plant breeders' demand for traits evolves. Until new wide crossing and molecular techniques substantially reduce the cost and time constraints on evaluation and pre-breeding, we should expect that collections of landraces and wild relatives will be seldom used. It makes economic sense to turn to landraces and wild relatives in situations where breeding lines have been searched extensively without success and the economic problem is large.

A third conclusion is that, even in large collections, non-trivial benefits may be associated with marginal accessions. We have not yet attempted to model the marginal value of accessions, but the results presented here are suggestive. There are some situations in which large searches are economically profitable, and, by extension, in which large collections will be valuable. The question that remains is *how often* large searches are warranted, and with what expected payoffs.

The results provided in this paper are illustrative. The main summary point is that it is possible—and straightforward—to arrive at empirically sensible answers to questions about the valuation and management of *ex situ* collections of genetic resources. To answer such questions, we propose a relatively simple framework based on search theory. Our analysis depends on a two simplifying assumptions, neither of which is particularly restrictive. First, we have modeled the use of an *ex situ* germplasm collection as a source of resistance to diseases and pests, rather than as a source of other traits of agronomic importance. In the past, breeders' demand for new sources of resistance to diseases and pests appears to have motivated most searches of *ex situ* wheat collections. Many experts believe, however, that future increases in yield potential in bread wheat, and hence the demand for genetic resources, are most likely to arise from quantitative characteristics that are more costly to evaluate (Reynolds, Rajaram, and McNab, 1996). Our model can be adapted for analyzing issues related to the search for such traits.

Our second major simplifying assumption is that scientists "know" the form of the distribution of useful traits across all relevant populations, when often they do not. In principle, however, our analysis would be essentially identical if scientists simply "guessed" the distributions on the basis of all available information. We do not, however, model the process by which scientists may *learn* about the distributions of useful traits.

4.7. IMPLICATIONS

The notion that unused gene banks are "seed morgues" and lack value is a misreading of the economics of search. Our model implies that, with current search and transfer techniques, collections of landraces will indeed be used only on rare occasions, but high values may be associated with those occasions. In other situations, it will be "efficient," in an economic sense, to keep landraces sitting unused in banks.

Further, it is important to recognize that we have used the word "utilization" in its narrowest sense. Even if breeding programs infrequently place direct demands on gene banks, each year large numbers of accessions are sent out on request to scientists for the purpose of genetics research and increasing the knowledge of the biochemical and molecular bases of certain traits. Gene banks are "utilized" for the accumulation of scientific knowledge, which in turn renders their use by breeding programs more strategic.

Several extensions to the analysis presented here bear mention. In our analysis, researchers make a single decision about how many materials to evaluate, and the resulting search is conducted at a moment in time. More realistically, however, we could allow for a sequenced search in which different numbers of materials are screened at different points in time. Sequential search would allow researchers to improve their prior information about distributions of resistance, to identify subpopulations with different resistance distributions, and to allow for repeated draws (Rausser and Small, 1997). Does it make sense to search initially through a small sample of genetic resources, as a way of assessing the likely differences across types of material, or as a way of acquiring specialized knowledge about where to find desirable traits?

Questions of this kind can be answered using modifications of the techniques and approaches outlined here. Some questions are harder. How, for example, will new technologies affect the materials that can be searched, and how will they affect the optimal search strategies? Will the current wave of biotechnological change raise or lower the marginal value of genetic resources? These are questions of policy importance that remain to be addressed in future research.

Acknowledgments

Among those who have contributed helpful comments and suggestions to the research described in this chapter are: N. Borlaug, C. Doss, J. Dubin, J. Ekboir, R. E. Evenson, P. W. Heisey, L. Gilchrist, M. T. Jackson, S. Lemarié, S. Rajaram, M. Reynolds, G. Traxler, and M. van Ginkel. H. Bockelman, Director of the US Department of Agriculture's National Small Grains Collection, made available part of the Genetic Resources Information Network database. W. Chang provided research assistance.

Notes

1 This chapter is based on an extended study described in Gollin, Smale, and Skovmand (1998).
2 Genetic Resources Information Network (GRIN) of the United States National Small Grains Collection, USDA, Agricultural Research Service.
3 Using the actual discrete distribution, a draw of size 152 breeding lines will have a 99% probability of achieving useful resistance. Using the smoothed beta distribution, a draw of only 95 breeding lines will have a 99% probability of achieving useful resistance. "Useful resistance" is defined to be a resistance score below 2.5 on a 10-point scale.

References

Autrique, E., M. M. Nachit., P. Monneveux, S. D. Tanksley, and M. E. Sorrells. 1996. Genetic diversity in durum wheat based on RFLPs, morphophysiological traits, and coefficient of parentage. *Crop Science* 36: 735–42.
Brown, G. Jr., 1990. Valuation of genetic resources. In G. H. Orians, G. M. Brown Jr., W. E. Kunin, and J. E. Swierzbinski (eds.), *The Preservation and Valuation of Biological Resources*. Seattle: University of Washington Press.
Byerlee, D., and P. Moya. 1993. *Impacts of International Wheat Breeding Research in the Developing World, 1966–90*. Mexico, D.F.: International Maize and Wheat Improvement Center (CIMMYT).
Chen, H. B., J. M. Martin, M. Lavin, and L. E. Talbert. 1994. Genetic diversity in hard red spring wheat based on sequence-tagged-site PCR markers. *Crop Science* 34: 1628–1632.
CIMMYT (International Maize and Wheat Improvement Center). 1985. *Wheat Producing Regions in Developing Countries*. Mexico, D.F.: CIMMYT.
du Toit, F. 1987. Resistance in wheat (*Triticum aestivum*) to *Diuraphis noxia* (Hemiptera: Aphididae). *Cereal Research Communications* 15: 175–179.
Dubin, H. J., and S. Rajaram. 1996. Breeding disease-resistant wheats for tropical highlands and lowlands. *Annual Review of Phytopathology* 34: 503–26.
Evenson, R. E. 1993. Genetic resources: Assessing economic value. Mimeo. New Haven: Yale University, Department of Economics.
Evenson, R. E., and Y. Kislev. 1975. *Agricultural Research and Productivity*. New Haven: Yale University Press.
FAO (Food and Agriculture Organization of the United Nations). 1996. *The State of the World's Plant Genetic Resources for Food and Agriculture*. Background documentation prepared for the International Technical Conference on Plant Genetic Resources, Leipzig, Germany, 17–23 June, 1996. Rome: FAO.
Frankel, O. H., and M. E. Soulé. 1981. *Conservation and Evolution*. Cambridge, UK: Cambridge University Press.
Fuentes, S., and L. Gilchrist. 1994. O programa de septoria de CIMMYT. *Melhoramento* 33: 507–523.
Gilchrist, L. I., and A. Mujeeb-Kazi. 1996. Septoria tritici leaf blotch resistant germplasm derived from bread wheat/D genome synthetic hexaploids. Poster presented at the Annual Meetings of the American Society of Agronomy, Indianapolis, August,1996.
Gilchrist, L. I., and B. Skovmand. 1995. Evaluation of emmer wheat (*Triticum dicoccon*) for resistance to *Septoria tritici*. In L. Gilchrist, M. van Ginkel, A. McNab, and G. H. J. Kema (eds.), *Proceedings of a Septoria tritici Workshop*. Mexico, D.F.: International Maize and Wheat Improvement Center (CIMMYT).
Gollin, D., M. Smale, and B. Skovmand. 1998. *Optimal Search in Ex Situ Collections of Wheat Genetic Resources*. CIMMYT Economics Working Paper 98-03. Mexico, D.F.: CIMMYT.

Guldager, P. 1975. *Ex situ* conservation stands in the tropics. In L. Roche (ed.), *Methodology of Conservation of Forest Genetic Resources*. Rome: Food and Agriculture Organization of the United Nations (FAO).

Great Plains Agricultural Council on Russian Wheat Aphid. 1990. Russian wheat aphid economic loss summary–1989. *Proceedings of the Fourth Russian Wheat Aphid Workshop*. Montana State University, Bozeman, Montana, 10–12 October, 1990.

Harvey, T. L. and T. J. Martin. 1990. Resistance to Russian wheat aphid, *Diuraphis noxia*, in wheat (*Triticum aestivum*). *Cereal Research Communications* 18: 127–129.

Marasas, C., P. Anandajayasekeram, V. Tolnay, D. Martella, J. Purchase, G. Prinsloo. 1997. *Socio-Economic Impact of the Russian wheat aphid Control Research Program*. Gaborone, Botswana: Southern African Center for Cooperation in Agricultural and Natural Resources Research and Training (SACCAR).

Maxted, N., B. V. Ford-Lloyd, and J. G. Hawkes. 1997. Complementary conservation strategies. In N. Maxted, B.V. Ford-Lloyd, and J.G. Hawkes (eds.), *Plant Conservation: The In Situ Approach*. London: Chapman and Hall.

Pearce, D. W., and D. Moran, in association with the Biodiversity Programme of the IUCN (the World Conservation Union). 1994. *The Economic Value of Biodiversity*. London: Earthscan.

Pecetti, L., and A. B. Damania. 1996. Geographic variation in tetraploid wheat (*Triticum turgidum* ssp. *turgidum convar.* Durum) landraces from two provinces in Ethiopia. *Genetic Resources and Crop Evolution* 43: 395–407.

Plucknett, D. L., N. J. H. Smith, J. T. Williams, and N. M. Anishetty. 1987. *Gene Banks and the World's Food*. Princeton: Princeton University Press.

Rajaram, S., and M. van Ginkel. 1996. *A Guide to the CIMMYT Bread Wheat Section*. Wheat Program Special Report No. 5. Mexico, D.F.: International Maize and Wheat Improvement Center (CIMMYT).

Rausser, G. C., and A. A. Small. 1997. *Bioprospecting with Prior Ecological Information*. Giannini Foundation Working Paper No. 819. Berkeley: University of California–Berkeley, Department of Agricultural and Resource Economics.

Reynolds, M. P., S. Rajaram, and A. McNab (eds.). 1996. *Increasing Yield Potential in Wheat: Breaking the Barriers*. Mexico, D.F.: International Maize and Wheat Improvement Center CIMMYT.

Robinson, J. 1994. *Identification and Characterization of Resistance to the Russian Wheat Aphid in Small-Grain Cereals: Investigations at CIMMYT, 1990–92*. CIMMYT Research Report No. 3. Mexico, D.F.: International Maize and Wheat Improvement Center (CIMMYT).

Robinson, J., and B. Skovmand. 1992. Evaluation of emmer wheat and other Triticeae for resistance to Russian wheat aphid. *Genetic Resources and Crop Evolution* 39: 159–163.

Sayre, K. D., R. P. Singh, J. Huerta-Espino, and S. Rajaram. 1998. Genetic progress in reducing losses to leaf rust in CIMMYT-derived Mexican spring wheat cultivars. *Crop Science* 38: 654–659.

Scharen, A. L., and Sanderson, F. R. 1985. Identification, distribution, and nomenclature of the *Septoria* species that attack cereals. In A. L. Scharen (ed.) *Septoria of cereals: Proceedings of a Workshop*. USDA-ARS Publication No. 12. Washington, D.C.: US Department of Agriculture.

Shiva, V. 1993. *Monocultures of the Mind*. London, New Jersey, and Penang: Zed Books and the Third World Network.

Simpson, R. D., and R. A. Sedjo. 1998. The value of genetic resource for use in agricultural improvement. In R. E. Evenson, D. Gollin, and V. Santaniello (eds.), *Agricultural Values of Plant Genetic Resources*. Wallingford: CAB International.

Simpson, R. D., R. A. Sedjo, and J. W. Reid. 1996. Valuing biodiversity for use in pharmaceutical research. *Journal of Political Economy* 104: 163–85.

Souza, E., C. M. Smith, D. J. Schotzko and R. S. Zemetra. 1991. Greenhouse evaluation of red winter wheats for resistance to Russian wheat aphid (*Diuraphis noxia*, Mordvilko). *Euphytica* 57: 221–225.

Swanson, T. 1995. The values of global biodiversity: The case of PGRFA. Paper presented at the Technical Consultation on Economic and Policy Research for Genetic Resource Conservation and Use, International Food Policy Research Institute, Washington, DC, 21–22 June, 1995. Manuscript. Cambridge University, Faculty of Economics.

Tanksley, S. D., and S. R. McCouch 1997. Seed banks and molecular maps: Unlocking genetic potential from the wild. *Science* 277: 1063–1066.

Tsegaye, S., H. C. Becker, and T. Tesemma. 1994. Isozyme variation in Ethiopian tetraploid wheat (*Triticum turgidum*) landrace agrotypes of different seed color groups. *Euphytica* 75: 143–147.

Wright, B. D. 1997. Crop genetic resource policy: The role of *ex situ* genebanks. *Australian Journal of Agricultural and Resource Economics* 41: 81–115.

5 OPTIMAL COLLECTION AND SEARCH FOR CROP GENETIC RESOURCES

R. E. Evenson and S. Lemarié

5.1. INTRODUCTION

Plant breeders exploit genetic resources in their search for improved genetic performance in cultivated crop species. Prior to the development of modern crop breeding techniques toward the end of the 19th century, farmers themselves selected their own varieties within individual, cultivated species. These farmers' varieties, often referred to as landraces, have been the source of almost all the modern crop varieties developed and diffused among farmers in both developed and developing countries. Since the development of wide-crossing techniques several decades ago, breeders have been able to exploit the genetic potential in wild species, or in uncultivated species that share the same genera as cultivated species. Now, with the development of modern biotechnology and genetic engineering, genetic materials from a broad range of sources can be incorporated into modern crop varieties.

Ex situ collections of genetic resources have been developed for most major crop species. These include landraces and wild species, as well as advanced breeding lines and obsolete cultivars. While many collections now appear to include a high proportion of the genetic material of landraces and wild species that is considered to be potentially collectable, none are considered complete (FAO, 1996). The economic questions associated with further collection and evaluation require that a means for valuing the genetic resources be developed.

In this chapter, we develop an overall model to determine the optimal strategy for collecting crop genetic resources and searching them for traits of economic value. The model demonstrates several important points about the optimal size of collections and the value of genetic resources. As in similar studies (Evenson and Kislev, 1976; Simpson, Sedjo, and Reid, 1996; Simpson and Sedjo, 1998; Rausser and Small, 1997; chapter 4), the framework for analysis is search theory. Our model differs from these studies in three important ways. First, we consider a two-stage model of crop breeding. In the first stage, genetic resources are collected in *ex situ* gene banks. In the second stage, breeders search for traits to be incorporated in variety development. Second, we model explicitly the search for multiple rather than single traits. Third, we introduce a geographic dimension into the distribution of source materials. Each geographic region corresponds to a cluster of genetic resources whose distribution can be quite different from that found in the other clusters.

In the first part of the chapter, we consider the search for a single trait. We show first that the optimal size of the collection is related to the distribution of trait values associated with genetic resources. The optimal number of accessions in a collection is not necessarily closely linked to the number of potential acquisitions of genetic resources because a subsample of potential acquisitions may replicate closely the complete distribution. This is essentially the result obtained in several other models of search for genetic resources (Simpson, Sedjo, and Reid, 1996). When genetic resources have a geographic dimension, we show that a two-stage collection strategy may be optimal: in the first stage, the most promising region (or regions) is identified, and in the second stage, an intensive search is carried out in this region.

The analysis with multiple traits is introduced in the second part of the chapter. We show that the optimal collection will be larger than implied by the single trait model, because conservation costs are shared among the different traits. When collections are from multiple regions within the multiple trait model, we show that collection strategies can be quite complex and that relatively large collections covering virtually all regions (ecological niches) may be optimal. The findings are illustrated with a numerical example.

5.2. THE SINGLE TRAIT CASE

In this section we illustrate the mechanics of searching for a single trait in a collection of genetic resources, using two alternative distributions of traits. We denote the cost of search as a constant C^e for each search (draw) from a distribution $f(x)$. Searching and evaluating N genetic resources will then cost NC^e.

5.2.1. Search When Trait Values Are Distributed Exponentially

Denote the maximum x in an evaluated set as z ($z = \max_{1 \le n \le N}[x_n]$). Assume that z_0 is the value achieved to date. Then technical progress from selection of N genetic resources is $\Delta z = z(N) - z_0$. Suppose further that the marginal value of one unit of z is constant at v. The expected profit from this search is then:

(5.1) $E\pi = E\Delta z \cdot v - N \cdot C^e = v \cdot Ez - v \cdot z_0 - N \cdot C^e$

Evenson and Kislev (1976) evaluated the problem when $f(x)$ is the exponential distribution:

(5.2)
$$\begin{cases} f(x) = \dfrac{1}{\lambda} \bullet e^{-x/\lambda} \\ F(x) = 1 \bullet e^{-x/\lambda} \end{cases}$$

The general form of the distribution of z (i.e., the order statistic) is:

(5.3.)
$$\begin{cases} H_N(z) = [F(z)]^N \\ h_N(z) = N[F(z)]^{N-1} f(z) \end{cases}$$

With the exponential distribution, using a logarithmic approximation, the expected outcome of search is then

(5.4) $\quad Ez(N) = N \int_0^{+\infty} z[F(z)]^{N-1} f(z) dz \approx \theta + \lambda \bullet \ln(N)$,

with the first-order condition

(5.5) $\quad \dfrac{\partial E\pi}{\partial N} = v \dfrac{\partial Ez(N)}{\partial N} - C^e = \dfrac{\lambda v}{N} - C^e = 0.$

The optimal level of search is:[1]

(5.6) $\quad N^* = \dfrac{\lambda v}{C^e}$,

and the search (research) is undertaken if the expected profit is positive. With the exponential distribution, given the conditions of optimal search in equation 5.6, the positive profit condition can be written as the minimum progress to make:

(5.7) $\quad Ez(N^*) \geq z_0 + \lambda$.

5.2.2. Search When $f(x)$ Is the Uniform Distribution

The logarithmic search function in equation 5.4 is a good approximation to order statistics with many distribution functions in addition to the exponential distribution, including the normal distribution, but may be less satisfactory when $f(x)$ is bounded. The expression of crop traits generally takes the form of a bounded distribution. Here, we examine the implications of its use in the case of the uniform distribution, a bounded distribution which is simple enough to be presented briefly .

Let x be uniformly distributed on $[a, a+d]$. The density function is $f(x) = 1/d$ when $a \leq x \leq a+d$ ($f(x) = 0$ otherwise) and the expected maximum trait value is:

(5.8) $\quad Ez(N) = \dfrac{d \bullet N}{N + 1}$

The marginal returns to selection are positive and decreasing, and the first-order condition leads to the following optimal investment in research when second-order conditions are also satisfied:

$$(5.9) \quad N^* = \sqrt{\frac{dv}{C^e}} - 1.$$

To summarize, the optimal investment in research (N^*) increases in both cases with v/C^e and the standard deviation of the distribution. When the logarithmic approximation is correct, N^* increases linearly with the two parameters. With the uniform distribution, N^* increases in the square root of the two parameters. The results are similar for the two distributions, and in the remainder of this chapter, we will use the logarithmic function.

5.2.3. Optimal Size of Collection for a Single Trait

The selection model shows that whatever the form of the distribution for x, with optimal decisionmaking, the profit from research increases with the variance of the distribution. We suppose now that this variance can be modified by collecting additional genetic resources and use a two-stage model to determine the optimal size of collection implied by the search for a single trait.

In Stage 1, a collection of N_1 accessions of genetic resources is made. The value x of each accession is drawn from a continuous distribution $f(x)$ and draws are made with replacement. The assumption of sampling with replacement may not realistic, but it simplifies the mathematics. The distribution is unknown in Stage 1. The cost of research in Stage 1 is $N_1 C^c$. C^c is the constant marginal cost of collecting and conserving an accession, cumulated over time.

In Stage 2, the collection is searched as described above. N_2 accessions are drawn randomly from the collection without replacement ($N_2 \leq N_1$) and evaluated. The maximum trait value from these N_2 draws gives the gains from search (research). The cost of research in Stage 2 is $N_2 C^e$, where C^e is the constant marginal cost of selection.

The expected gains of research can now be written as $Ez(N_1, N_2)$. As long as $N_2 \leq N_1$, $Ez(N_1, N_2) = Ez(N_2)$. This is true because it is equivalent to draw a random subsample (without replacement) of N_2 plants in a population of N_1 plants made by random draw in a given distribution, and to draw directly the N_2 plants in the same distribution.

The expected profit from research is:

$$(5.10) \qquad E\pi(N_1, N_2) = vEz(N_1, N_2) - N_1 C^c - N_2 C^e$$

and profit is maximized using backward induction. First, the optimal selection strategy in Stage 2 is computed for a given value of N_1, with the following first-order conditions:

$$(5.11) \qquad \frac{\partial E\pi(N_1, N_2)}{\partial N_2} = v \cdot \frac{\partial Ez(N_1, N_2)}{\partial N_2} - C^e$$

Second-order conditions are satisfied. Using equation 5.4 and imposing the constraint that $N_2 \leq N_1$ leads to the optimal selection strategy:

$$(5.12) \quad N_2^* = \min(N_1, N_2^0) \text{ with } N_2^0 = \frac{\lambda v}{C^e}.$$

Next, the optimal strategy for collection in Stage 1 can be found given that N_2 is chosen optimally. If $N_1 \leq N_2^0$ then $N^*_2 = N_1$, and $E\pi(N_1, N_2) = v\, Ez(N_1) - N_1\, (C^c + C^e)$. The solution is then

$$(5.13) \quad N^*_1 = \frac{\lambda v}{C^c + C^e}.$$

If $N_1 > N_2^0$ then $N^*_2 = N_2^0$, and $E\pi(N_1, N_2^0) = v\, Ez(N_2^0) - N_1\, C^c - N_2^0 C^e$. This function is linear and strictly decreasing in N_1. Consequently, the profit when $N_1 > N_2^0$ is always less than the maximal profit when $N_1 \leq N_2^0$. Finally, the unique optimal strategy when maximizing on two periods is:

$$(5.14) \quad N^*_1 = N^*_2 = \frac{\lambda v}{C^c + C^e}.$$

Figure 5.1 provides an illustration of this maximization exercise. When N_1 is given, the value of N_2 which maximizes the second period profit is $N_2 = \min(N_1, N^0_2)$ and the second period profit with optimal N_2 is non-decreasing in N_1. In the two-stage model, profits in the second period are reduced by the conservation cost and the new curve is maximum for $N_1 = N^*_1 \leq N^0_2$. Finally, the optimal collection size *when maximizing for one trait* is the optimal research strategy minus "something" which will be more important if conservation cost is high. Above this level, the value of the marginal accession of genetic resources is low. In this respect, the two-stage search model for a single trait yields results similar to those shown by Simpson, Sedjo, and Reid (1996).

5.2.4. Implications of Regional Distributions of a Single Trait
Suppose it is possible to identify k different geographical regions over which the single trait is distributed. Trait values in each region can be described with a density function f_k. Suppose further that the economic value of the trait (v) is independent of regions. Now it may be possible to economize further in the first stage by collecting accessions in only a few (or one) regions. In particular, suppose we search for traits by drawing a subsample, N_{2k}, from each of N_{1k} accessions. The strategy for the assembly and the exploitation of a collection can be represented as a set of couples $\{(N_{1k}, N_{2k})\}$, each couple corresponding to the assembly and the exploitation of a subcollection. For

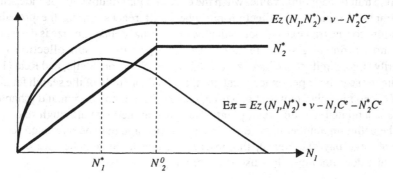

Figure 5.1. Stage 2 and Two-Stage Profit When N₂ Is Chosen Optimally.

example, each region of crop domestication and diversity might represent a subcollection. One strategy would be to collect only in region k and exploit only this subcollection in searching for traits of economic value. Such a corner strategy is denoted (N_{1k}, N_{2k}). The expected profit in any particular region with a corner strategy is:

$$(5.15) \qquad E\pi\{(N_{1k}, N_{2k})\} = v \cdot \max[(N_{1k}, N_{2k})] - \sum_k [N_{1k} C^c_k + N_{2k} C^e]$$

In general, given any research strategy, there is a special region k^* which provides the greatest economic gains.

$$(5.16) \qquad Ez_{k^*}(N_{1k^*}, N_{2k^*}) = \max_k [Ez_k(N_{1k}, N_{2k})]$$

Moreover, we have the following inequality:

$$(5.17) \qquad N_{1k^*} C^c_{k^*} + N_{2k^*} C^e \le \sum_k [N_{1k} C^c_k + N_{2k} C^e]$$

The sum of these two last equations leads to the conclusion that $E\pi\{(N_{1k}, N_{2k})\} \le E\pi(N_{1k^*}, N_{2k^*})$. In other words, any strategy is dominated by a corner strategy. The optimization problem is straightforward. We need only to solve the problem in each region according to section 5.2.3., and then compare K different profit levels in order to select the corner stra̕ᵗgy that generates the maximum economic gains.

If the assumption of perfect information on f_k is relaxed, or in the presence of risk, or if there are technical or cost complementarities in the search, the corner solution can be lost and the exploitation of more than one region may be optimal. One way to consider this problem is to suppose that information on the parameters of f_k can be obtained during a preliminary period (period 0). The strategy in the first period would be to collect a small sample in each of the regions and estimate the parameters of f_k from this sample. The cost of such a strategy is fixed, so there is no effect on the optimal collections. If perfect (or good enough) information can be obtained this way, the optimization is solved as described and a corner solution is obtained. If the information revealed during period 0 is poor, then there may be a research interest in rejecting the corner solution and continuing the learning process.

In the first section of this chapter, we have shown that in the collection and search for a single trait of economic value, when the expected maximum value is independent of geographical regions of collection but where differences among the probability distributions for the trait exist geographically, the optimal collection size is determined by the most promising subcollection. The most promising subcollection is not necessarily large. Similar conclusions were reached by Simpson, Sedjo, and Reid (1996) regarding the search for pharmaceutical products. In their study of the search for single traits of economic value in wheat breeding, Gollin, Smale, and Skovmand (chapter 4) have made a related, but different, point. They have argued that although searches in which the optimum number of accessions screened is large may be rare, such searches may have large payoffs. Hence, the observation that large searches are conducted infrequently does not imply low use value of marginal accessions.

Neither of these models, however, nor the model presented in the first section of the chapter, implies that the collection size (N_1) should be larger than the search size (N_2). This implication is changed when multiple trait searches are considered in the next section. Note that collection strategies are based on collecting and evaluating genetic resources in the form of actual plants. Each accession then contains a great deal of genetic information that may be of value with respect to many traits. Furthermore, crop breeders can generally select for a single valuable trait and eliminate (at a cost) traits of low value associated with an accession.

5.3. OPTIMAL COLLECTION AND SELECTION FOR MULTIPLE TRAITS

5.3.1. Decomposition with Several Simple Traits

The value of a genetic resource is related to multiple uses for different regions of production and consumption. In this section, we extend the model presented above to the search for multiple traits in crop breeding. Each trait is indexed by j ($j \in \{1,...,J\}$) and the (constant) marginal value of the trait is v_j. The technical value of a plant population n for trait j is x_{jn}, and its economic value is:

$$(5.18) \qquad V_n = \sum_{j=1}^{J} v_j x_{jn}$$

In this equation, the different traits are assumed to be independent in an economic sense. Conceptually, the value of plant n for trait j will be the product $v_j x_{jn}$, whatever the value of the plant for the other traits. In practice, disaggregating and measuring that value is difficult. In terms of the model we have developed above to depict the collection of genetic resources and search for traits, the economic value of the plant population is the sum of the maximum values achieved for each trait.

$$(5.19) \qquad V = \sum_{j=1}^{J} \left[v_{jk} \max_{1 \leq n \leq N}[x_{jn}] \right] = \sum_{j=1}^{J} [v_j z_k]$$

5.3.2. The Maximization Problem with Multiple Traits

Consider again the two-stage model developed in the first section of this chapter. In Stage 1, collection is undertaken in K different regions. N_{1k} is the number of accessions collected in region k. We assume that the information on the donor region is stored for each accession, so the total collection can be represented as a series of K subcollections, each of these corresponding to its geographical origin. The collection is exploited and leads to technical progress on J simple traits. N_{2jk} is the number of accessions which are evaluated in the subcollection k for the trait j. The distribution of trait values within each region is assumed to be known and is represented with the density function f_{jk}. The general form of the expected profit is:

$$(5.20) \qquad E\pi = \sum_j v_j \max_k \left[E z_{jk} (N_{1k} N_{2jk}) \right] - \sum_k N_{1k} C_k^c - \sum_k \sum_j N_{2jk} C_j^e$$

A complete strategy is a set $\{(N_{1k}, N_{2jk}, ..., N_{2Jk})\}$ which contains $K(J+1)$ choice variables.

5.3.3. Simplified Version of General Problem

For any given trait, there is no reason to exploit more than one subcollection. The optimal strategy will be a corner solution. For any trait j, the geographical region that provides the best genetic progress, k^*_j, can be identified by solving

$$(5.21) \qquad \max_k \left[Ez_{jk}(N_{1k}N_{2jk}) \right] = Ez_{1k^*_j}(N_{1k^*_j}, N_{2jk^*_j}),$$

so that any original strategy is replaced by a corner strategy that provides the greatest profit. Given the corner solution, the subcollection k^*_j can be exploited at minimum cost:

$$(5.22) \qquad \max_k \left[Ez_{jk}(N_{1k}, N_{2jk}) \right] - \sum_k N_{2jk} C^e_j \le Ez_{jk^*_j}\left(N_{1k^*_j}, N_{2jk^*_j}\right) - N_{2jk^*_j} C^e_j,$$

Now consider strategies in which exploitation of each trait is made in only one region. An example of such a strategy is given in Table 5.2. Define Ω_k as the set of traits selected from region k (in Table 5.2, $\Omega_1 = \{1,2\}$, $\Omega_2 = \{3\}$, $\Omega_3 = \{4,5\}$). By construction, $\{\Omega_k\}$ (the set of all sets Ω_k) is a partition of the set of all traits $\{1,...,J\}$. Each trait is selected in one and only one subcollection. A strategy has three elements :

$$(5.23) \quad \begin{vmatrix} \{\Omega_k\} & \text{the partition of } \{1,...,J\} \text{ which indicates the subcollection to exploit} \\ & \text{for each trait} \\ \{N_{1k}\} & \text{the number of accessions to collect in each region} \\ \{N_{2j}\} & \text{the choice variable for exploiting each trait} \end{vmatrix}$$

The expected profit is

$$(5.24) \qquad E\pi = \sum_k \sum_{j \in \Omega_k} v_j Ez_{jk}(N_{1k}, N_{2j}) - \sum_k N_{1k} C^c_k - \sum_j N_{2j} C^e_j,$$

which can be rewritten as the sum of profit made for each subcollection:

$$(5.25) \qquad E\pi = \sum_k E\pi_k \quad \text{with: } E\pi_k = \sum_{j \in \Omega_k} v_j Ez_{jk}(N_{1k}, N_{2j}) - N_{1k} C^c_k - \sum_{j \in \Omega_k} N_{2j} C^e_j.$$

Because of the partition of $\{1,...,J\}$ by $\{\Omega_k\}$, the expected profits of the different subcollections are independent from each other and they can be maximized independently, leading to an optimal subcollection strategy $(N^*_{1k}, \{N^*_{2j}\}_{j \in \Omega_k})$. We can define the union of the different optimal subcollection strategies as the all-collection optimal strategy with a given partition $\{\Omega_k\}$. Finally, by comparing the optimal profit for all the possible partitions, we identify the complete, all-collection, optimal strategy.

5.3.4. Determining the Optimal Size of Collection

The properties of the optimal strategy can be compared with the solution of J independent maximization problems, each one corresponding to the collection and search for one particular trait. The set $\{\Omega_k\}$ defines, for each trait, the region where the collection has to be done. This set is considered as given, and the final step is to compare the solutions for all the possible sets.

The sign $\sim$ is used to distinguish the variables of the profits of the independent problems from the variables of the joint profit maximization problem. The expected profit of one problem (corresponding to the trait j) is:

$$(5.26) \qquad E\tilde{\pi}_j = v_j Ez_{jk}(N_{1k}, N_{2j}) - N_{1k}C_k^c - N_{2j}C_j^e$$

The solutions from J maximization problems can be described in terms of three sets:

$\{\Omega_k\}$ the partition of $\{1,...,J\}$ which indicates the subcollection to exploit for each trait

$\{\tilde{N}^*_{1j}\}$ the number of accessions to collect the best region for trait j (corner solution)

$\{\tilde{N}^*_{2j}\}$ the choice variable for exploiting each trait

This solution is not yet fully comparable to the solution of the complete program described before because the choice variables for the first period are subscripted here by the trait number j. However, the variable $\tilde{N}^*_{1k} = \max_{j \in wk}[\tilde{N}^*_{1j}]$ can be defined.

The optimal solutions of the J independent problems are then used to build the following reference strategy:

$$(5.27) \qquad \begin{vmatrix} \{\Omega_k\} \\ \{\tilde{N}^*_{1k}\} \text{ with } \tilde{N}^*_{1k} = \max_{j \in \Omega k}[\tilde{N}^*_{1j}] \\ \{\tilde{N}^*_{2j}\} \text{ the choice variable for exploiting each trait} \end{vmatrix}$$

This strategy is similar to the simplified strategy of the joint profit maximization: (1) the partition $\{\Omega_k\}$ is identical by construction; (2) for each trait, the exploitation is only a corner solution; (3) when we aggregate over the different traits, collection in different regions is of interest because each region provides the best of a type of genetic resource.

Once this reference strategy has been developed, we can define two types of traits within each set Ω_k. A major trait is a trait for which the optimal number of populations to collect is maximum, or j such that $\tilde{N}^*_{1j} = \tilde{N}^*_{1k}$. If there are several major traits, we choose the trait with the smallest index j arbitrarily, without any influence on the results. In Table 5.2, the major traits are: (1) trait 1 for region 1; (2) trait 2 for region 2; (3) trait 4 for region 3. The minor traits are all the other traits of Ω_k.

The profit in region k can then written as the sum of profits defined at the trait level ($E\pi_k = \Sigma_{j \in \Omega k} E\pi_j$):

$$(5.28.) \begin{cases} E\pi_k = v_j Ez_{jk}(N_{1k}, N_{2j}) - N_{1k}C_k^c - N_{2j}C_j^e = E\tilde{\pi}_j & \text{if } j \text{ is the major trait of } \Omega_k \\ E\pi_k = v_j Ez_{jk}(N_{1k}, N_{2j}) - N_{2j}C_j^e & \text{if } j \text{ is a minor trait of } \Omega_k \end{cases}$$

Figure 5.2 provides an illustration of profit function at the trait level, when N_{2j} is chosen optimally. A comparison can be made with the curves of Figure 5.1, in which the second-stage and two-stage profit functions for one trait in one region are depicted.

The profit with major trait a is similar to the two-stage profit, since it increases and then decreases. The profit with minor traits (b and c) resembles the second-stage profit in Figure 5.1, where the costs of conservation have not been included.

5.3.5. The Sign of the Marginal Joint Profit

The marginal profit at the region level can be decomposed as the sum of marginal profits at the trait level:

$$(5.29) \qquad \frac{\partial E\pi_k}{\partial N_{1k}} = \sum_{j \in \Omega_k} \frac{\partial E\pi_j}{\partial N_{1k}}$$

The sign of each marginal profit at the trait level is known for $N^*_{1k} = \tilde{N}^*_{1k}$. For the major trait (e.g., trait a in Figure 5.2), we have $\partial E\pi_j/\partial N_{1k} = 0$ because $\tilde{N}^*_{1k} = \tilde{N}^*_{1j}$. For the minor traits, we need to consider the optimal level of exploitation N^0_{2j}. Two cases have to be distinguished: (1) if $N^*_{1k} > N^0_{2j}$ (e.g., trait b in Figure 5.2), then the optimal exploitation is $N^*_{2j} = N^0_{2j}$ whatever N^*_{1k}, and consequently, $\partial E\pi_j/\partial N_{1k} = 0$; (2) if $N^*_{1k} \leq N^0_{2j}$ (e.g., trait c in Figure 5.2), then the optimal exploitation is $N^*_{2j} = N_{1k}$, and $\partial E\pi_j/\partial N_{1k} > 0$.

Finally, if we can find at least one minor trait such that $N^0_{2j} > \tilde{N}^*_{1k}$, then:

$$(5.30) \qquad \left. \frac{\partial E\pi_k}{\partial N_{1k}} \right|_{N_{1k} = \tilde{N}^*_{1k}} > 0 \Rightarrow N^*_{1k} = \tilde{N}^*_{1k}$$

Otherwise, $N^*_{1k} = \tilde{N}^*_{1k}$.

Using Figure 5.2, if we consider only the traits a and b, then no minor trait satisfies the condition $N^0_{2j} > \tilde{N}^*_1$, and $N^*_{1k} = \tilde{N}^*_{1k}$. If we consider the three traits (a, b, and c), then the minor trait c satisfies the condition $N^0_{2c} > \tilde{N}^*_1$, and $N^*_{1k} > \tilde{N}^*_{1k}$.

Starting from $\tilde{N}^*_{1k}$ and increasing N_{1k}, two changes occur. For the major trait, the marginal profit becomes negative because above the optimum value, $\partial E\pi_j/\partial N_{1k} < 0$. For the minor traits that satisfy $N^0_{2j} > \tilde{N}^*_{1k}$, the marginal profit decreases until the level $N^*_{1k} = N^0_{2j}$, and thereafter the marginal profit is nil. Because of these two changes, the subcollection marginal profit $\partial E\pi_k/\partial N_{1k}$ decreases and becomes negative at some point, so an optimal subcollection size can be determined.

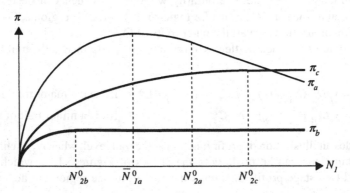

Figure 5.2. An Illustration of the Trait-Level Profit Function (3 Traits).

5.3.6. Synthesis

Using the framework of search theory, we have outlined a method for determining the optimal size of *ex situ* collections.[2] First, we find the optimal size of the subcollection corresponding to region k (with a given partition $\{\Omega_k\}$), in three steps:

1. For each trait in Ω_k, problems are considered independently and the optimal sizes $\tilde{N}^*_{1k}$ and N^0_{2j} are identified.
2. The traits are ranked. Let the major trait be a and the minor traits be arranged in order of decreasing values as $N^0_{2b} > N^0_{2c} > \ldots$
3. A search for the optimal subcollection size N^*_{1k} can then be undertaken. By construction, we know that the value is greater or equal to $\tilde{N}^*_{1k}$. If $N^0_{2b} < \tilde{N}^*_{1k}$, then all the minor traits have zero marginal profit above $\tilde{N}^*_{1k}$, and $N^*_{1k} = \tilde{N}^*_{1k}$. In the converse case, we can calculate the marginal profit $(\partial E\pi_k/\partial N_{1k})$ at the different levels N^0_{2j} starting from b (the higher level of N^0_{2j}) until we find a positive marginal profit. We then have a range delimited by two values of N^0_{2j}, within which the marginal profit has a simple form. The equation $\partial E\pi_k/\partial N_{1k} = 0$ can then be solved analytically.

Each of the three steps can be repeated for every region and lead to an optimal size of the total collection for a given partition $\{\Omega_k\}$. The process can then be repeated for every possible partition. The number of possible partitions is the number of possible associations of one of the K different regions to each of the J different traits. The maximum number is then K^J. However, the real number is generally smaller because there are some traits for which the number of possible regions is less than K. For example, when considering the parameters of Table 5.1, we know that only collection in region 1 for trait 1 $(\lambda_{12} = \lambda_{13} = 0)$ is of interest. For this case, 12 different partitions are possible $(1 \times 2 \times 3 \times 2 \times 1)$.

5.4. AN ILLUSTRATION WITH ALTERNATIVE SEARCH COSTS IN ONE REGION

Table 5.1 summarizes parameters associated with several traits. For each trait j, a marginal value (v_j) and search cost (C^e_j) are given. Conservation or collection costs (C^c_k) are provided for each region. Search parameters from the underlying distribution $(\lambda$, equation 5.4) are specified by region and trait.

Table 5.1. Parameter Values for Tables 5.2 and 5.3

	Trait number	$J=1$	2	3	4	5
	Marginal value ($\times 10^3$)	$v_1=100$	50	50	10	10
	Search cost	$C^e_1=300$	150	200	100	100
Region (k)	Conservation cost	Value of λ				
$k=1$	$C^c_1=300$	$\lambda_{11}=0.75$	1.00	0.75	0.00	0.00
2	400 or 800	0.00	0.75	1.00	0.75	0.00
3	250	0.00	0.00	0.75	1.00	0.75

Table 5.2 reports an illustration of optimal search N^*_{2j} by trait and region, and optimal collection N^*_{1j} by regions when collection costs (C^c_2) in region 2 are 400. The left-hand side of Table 5.2 reports the results when searches for traits are independent. In this case, recall that $N^*_{1j} = N^*_{2j}$, meaning that collections are the same size as searches. Note that the size of the optimal collection varies by trait, but that if all five traits were sought, the total collection size would be 237, including 125 from region 1, 83 from region 2, and 29 from region 3. Total collection size is therefore higher than in any single trait search.

When joint profit maximization is undertaken as noted in the right-hand side of Table 5.2, there are economies to collecting for more than one trait. Collection size increases to 289 (167 from region 1, 83 from region 2, and 39 from region 3). Searches are constrained by the λ's in Table 5.1. This finding illustrates the basic point of the multiple trait model: optimal collection size in the search for multiple traits is greater when profits are jointly maximized than is implied by the sum of independent maximization problems for single traits. The same holds true for profits (732 as compared to 682).

Table 5.3 illustrates the effect of a rise in the cost of collecting and conserving in region 2 from 400 to 800. With independent maximization of profits, optimal collection size is reduced in region 2 and profits associated with trait 3 are lower. When profits are maximized jointly, the optimal collection size from region 2 diminishes to zero, while collection size from region 1 increases because trait 3 can be more efficiently obtained from region 1 (given the λ parameters postulated).

Table 5.2. An Illustration of Independent and Joint Profit Maximization (case 1: C^c_2=400)

Region	Independent profit maximization						Joint profit maximization						
	$\tilde{N}^*_{1k}$ $j=1$	2	3	4	5	Maximum	N^*_{1k}	N^*_{1k} $j=1$	2	3	4	5	Profit ($\times 10^3$)
1	125	111	0	0	0	125	167	167	167	0	0	0	514
2	0	0	83	0	0	83	83	0	0	83	0	0	171
3	0	0	0	29	21	29	39	0	0	0	39	39	47
Profit ($\times 10^3$)	287	186	171	24	15								
Total	682												732

Table 5.3. An Illustration of Independent and Joint Profit Maximization (case 2: C^c_2=800)

Region	Independent profit maximization						Joint profit maximization						
	$\tilde{N}^*_{1k}$ $j=1$	2	3	4	5	Maximum	N^*_{1k}	N^*_{1k} $j=1$	2	3	4	5	Profit ($\times 10^3$)
1	125	111	0	0	0	125	171	171	171	171	0	0	673
2	0	0	50	0	0	50	0	0	0	0	0	0	0
3	0	0	0	29	21	29	39	0	0	0	39	39	47
Profit ($\times 10^3$)	287	186	145	24	15								
Total	657												720

This case illustrates another feature of genetic resource conservation. When alternative (substitute) resources exist, collection costs can lead to shifts in sources by regions. If a small region is a relatively rich (high λ) source for a particular trait, collection costs C^c may be low, and marginal values may be high. It will always pay to collect from such a region when profits are maximized independently and will almost always pay to do so even when they are maximized jointly. If there is an alternative source for the trait, a rise in collection costs could lead to abandoning collection activities in the richer region in favor of searching in a less rich (i.e., lower λ) region.

5.5. CONCLUSION

The basis for this chapter is a two-stage model with a collection activity in the first stage and a selection of the best accessions within this collection in the second stage. This model provides similar results compared to other studies, when applied to similar cases such as the search for a single trait, in a single period, or in a single region. However, when breeders collect in different regions for multiple traits, the model has different implications.

First, we show that the optimal size of a collection is highly sensitive to the number of traits for which the collection is made. The collection size needed to satisfy the search for multiple and complex traits will necessarily be higher than that implied by the successive, independent searches for single traits. A model depicting the joint search for multiple traits clearly conforms more realistically to the general case of crop improvement in a breeding program than a model of search for a single trait. Models of search for single (simply inherited or complex) traits reflect the way in which crop breeding programs may resolve particular, well-defined problems, such as those found with some plant diseases and pests.

Second, we show that the optimal size of collection is sensitive to the diversity of the distribution functions among the different regions providing source materials. The optimal size of collection will be larger if traits are clustered in "niches" rather than distributed randomly across the population. It is quite possible that optimal collection will entail sampling in virtually all regions because each new site may provide new genetic resources for new traits. We have developed simple illustrations of optimal collections to show these points.

In this chapter, we explicitly consider simply inherited traits. However, in most of the major crops, several traits of economic interest (yield, plant height, and others) have a much more complex nature. Complex traits are determined by several genes, while simple traits are determined by only one gene. By crossing complementary plants, it is possible to cumulate the best genes from both parents and create new plants with better combinations than any single parent has (Gallais, 1989). In its simplest form, the modeling of a complex trait is identical to the modeling of multiple simple traits in this chapter and leads to similar results.[3] In other words, instead of considering collections for simple traits in different regions, we can also consider a collection for only one complex trait and show that it will pay to collect in different regions, because each of these sites may provide good genetic resources for one particular locus influencing the complex trait.

We have developed the model assuming a constant marginal cost of collecting, conserving, and selecting genetic resources. Cost complementarities among regions, or increasing marginal costs, might affect the optimal solutions implied by the model. While we have motivated the work on multiple traits by referring to the potential value of recombining multiple traits in plant breeding, future research may investigate more explicitly the role of option values in the context of the model presented here. The model has implications for genetic resource conservation and management, which may be more fully developed through empirical application.

Acknowledgments

The authors thank Zhijie Xiao for his help on the development of the analysis with uniform distribution, and Michel Trommetter and Douglas Gollin for comments.

Notes

1 The second-order condition can be checked:

$$\frac{\partial^2 E\pi}{\partial N^2} = v \frac{\partial^2 Ez(N)}{\partial N^2} = -\frac{\lambda v}{N^2} < 0$$

2 The different steps described here are included in a computer program using C programming language. This program can be provided by the authors on request.
3 Appendix on this generalization can be provided by the authors.

References

Evenson, R. E., and Y. Kislev. 1976. A stochastic model of applied research. *Journal of Political Economy* 84 (2): 265–281.

FAO (Food and Agriculture Organization of the United Nations). 1996. Report on the state of the world's crop genetic resources for food and agriculture. Prepared for the Fourth International Technical Conference on Plant Genetic Resources, held in Leipzig, Germany, June 17–23, 1996. Rome: FAO.

Gallais, A. 1989. *Théorie de la Sélection en Amélioration des Plantes*. Paris: Masson.

Rausser, G. C., and A. A. Small. 1997. *Bioprospecting with Prior Ecological Information*. Giannini Foundation Working Paper 819. Berkeley: Department of Agricultural and Resource Economics, University of California, Berkeley.

Simpson, R. D., and R. A. Sedjo. 1998. The value of genetic resource for use in agricultural Improvement. In R. E. Evenson, D. Gollin, and V. Santaniello (eds.), *Agricultural Values of Plant Genetic Resources*. Wallingford: CAB International.

Simpson, R. D., R. A. Sedjo, and J. W. Reid. 1996. Valuing biodiversity for use in pharmaceutical research. *Journal of Political Economy* 104 (1): 163–185.

Conserving Crop Diversity on Farms

6 FARMERS' PERCEPTIONS OF VARIETAL DIVERSITY: IMPLICATIONS FOR ON-FARM CONSERVATION OF RICE

M. R. Bellon,[1] J.-L. Pham, L. S. Sebastian,
S. R. Francisco, G. C. Loresto, D. Erasga, P. Sanchez,
M. Calibo, G. Abrigo, and S. Quilloy

6.1. INTRODUCTION

There is growing interest in on-farm conservation as a complementary strategy for conserving rice genetic resources (Balakrishna, 1996; Bellon, Pham, and Jackson, 1997; Vaughan and Chang, 1992). On-farm conservation has been defined as farmers' continued cultivation and management of a diverse set of crop populations in the agroecosystems where the crop has evolved (Bellon, Pham, and Jackson, 1997). Compared to conservation in gene banks, on-farm conservation is dynamic, because the varieties that farmers manage continue to evolve in response to selection pressures. On-farm conservation also emphasizes the role of farmers in two ways. First, crops are not only the result of natural selection, but also of human selection and management. Second, farmers' decisions determine whether these populations are maintained. On-farm conservation depends on the active participation of farmers, and therefore it should be based on farmers' reasons and incentives to maintain diversity.

Four distinct ecosystems have been identified for rice production (Khush, 1984): irrigated, rainfed lowland, upland, and flood-prone ecosystems. These ecosystems represent different agroecological conditions and levels of agricultural intensification. Little is known, however, about the relative varietal and genetic diversity present in each, or about the levels of genetic erosion that may have occurred. The commonly accepted view is that the irrigated ecosystem is most likely to have the lowest level of varietal and genetic diversity, while the upland and flood-prone ecosystems have the highest, and the rainfed lowland ecosystem has an intermediate level. This view is based on two assumptions. The first is that there is an inverse relationship between the adoption of modern varieties and the levels of varietal and genetic diversity present in an ecosystem. The second is that there is a direct relationship between the maintenance of traditional varieties and the levels of varietal and genetic diversity found in an ecosystem.

Modern varieties have been widely adopted in the irrigated ecosystem, and a few of them may cover very large areas (Byerlee, 1994; David and Otsuka, 1994). There has been almost no adoption in the upland and flood-prone ecosystems, where traditional varieties still predominate (Chang, 1994; IRRI, 1993a, 1994). The rainfed lowland ecosystem presents an intermediate situation (IRRI, 1992).

The thesis of this chapter is that by studying patterns of adoption and loss of modern and traditional varieties across ecosystems, researchers may gain important insights into the opportunities to maintain varietal and genetic diversity on farms. Our purpose is twofold. First, we present a comparative study of farmers' perceptions of the varieties available to them, as well as the varietal and genetic diversity that farmers maintain across three rice ecosystems in the Philippines: upland, rainfed lowland, and irrigated ecosystems. Second, we explore the implications of these results for on-farm conservation of genetic resources.

This chapter presents several key findings. Modern and traditional varieties coexist in the upland and rainfed lowland ecosystems, while traditional varieties have completely disappeared from the irrigated ecosystem. The patterns of varietal and genetic diversity correspond to those predicted above, and they reflect the numerous concerns or characteristics of interest that rice farmers consider when choosing varieties to plant. Farmers perceive that their varieties perform differently with respect to these concerns, and can assess these differences. Their criteria for choosing or discarding varieties depends on the ecosystem. This study suggests that farmers' knowledge can provide several insights that are essential to on-farm conservation of genetic diversity. By eliciting farmers' knowledge, we can identify promising varieties or groups of varieties based both on desirable traits as well as their potential contribution to the genetic diversity of an agroecosystem. We can also identify the opportunity costs that may lead to the elimination of these varieties and look for ways to decrease these costs. In our view, the goal of on-farm conservation is to increase the likelihood that farmers will continue to grow varieties identified as important contributors to genetic diversity in the agroecosystem. To identify these varieties we combine farmers' evaluations with genetic data measured at the molecular level.

6.2. STUDY SITE AND METHODS

Cagayan Province, where the three rice ecosystems were studied, is located in northeastern Luzon, Philippines. This province was selected for several reasons: it includes the three rice ecosystems of interest for this research; farmers grow both modern and traditional rice varieties; collections of rice germplasm made in this area provided useful background information for the study; and the province is ethnically diverse (including Ilokano, Ibanag, Itawis, Ibaloys, and others). Table 6.1 presents key socioeconomic indicators of the communities that were chosen to represent each of the ecosystems.

Three municipalities in Cagayan Province were selected as representative of the three rice ecosystems studied. Baggao represented the upland ecosystem, Iguig the rainfed lowland ecosystem, and Gattaran the irrigated ecosystem. Within each municipality, four communities (*barangay*) were chosen so as to include the greatest possible variation in the ecosystem for agroecological environment, infrastructure, and socioeconomic and ethnic conditions. In each *barangay*, four key informants were selected, both men and women, to represent the best recognized knowledge and ability to manage rice varietal diversity in that *barangay*.

In each *barangay*, two questionnaires were used. One, addressed to local authorities, focused on *barangay* characteristics, such as population, land use, and infrastructure. The second, answered by the key informants, dealt with farmers' management of varietal diversity. In this questionnaire we elicited information on varieties grown and discarded, sources and exchanges of seed, seed selection methods, and criteria used in choosing varieties. Questions relating to the socioeconomic characteristics of the informants, such as age, education, and land holdings, were also included.

Seed was collected from the key informants in every *barangay* to assess the actual genetic diversity of rice grown in the three ecosystems. The collection was done under the auspices of PhilRice, the national research institution for rice genetic resources. The genetic diversity of the accessions was studied with isozyme electrophoresis. The specific methods used and the data analysis procedures are presented in more detail in Bellon *et al.* (1998). In this chapter, we include only the genetic results that are relevant to the present analysis.

There are important differences between the communities representing each ecosystem. The rainfed lowland communities are located next to the highway, and good transportation links them to the provincial capital, Tuguegarao. The upland communities are isolated. The irrigated communities are closer to the main highway, but further away from Tuguegarao. The population and the number of households increase from the upland to the irrigated communities (Table 6.1). The highest population is in the rainfed lowland ecosystem, with the lowest in the upland ecosystem. The percentage of households engaged in farming was much lower in the rainfed lowland

Table 6.1. Key Socioeconomic Characteristics of the Study Area, by Ecosystem

Characteristic	Upland ecosystem	Rainfed lowland ecosystem	Irrigated ecosystem
Number of communities	4	4	4
Population (persons)	2,811	3,925	4,805
Population density (persons/ha)	0.13	4.36	2.51
Number of households	505	670	1043
Households engaged in farming (%)	98.6	56.3	80.8
Number of stores	15	23	77
Communities with electricity (%)	0	50	75

ecosystem, suggesting that off-farm activities are an important component of the local economy. There were also differences in the availability of infrastructure (e.g., electricity) and the degree of market integration, as indicated by the number of stores.

6.3. FARMERS' VARIETIES AND GENETIC DIVERSITY

The varieties identified by the farmers in each of the three ecosystems were classified as modern[2] or traditional and non-glutinous or glutinous, forming four groups: non-glutinous modern varieties; non-glutinous traditional varieties; glutinous modern varieties; and glutinous traditional varieties. Glutinous varieties were classified separately because they met a special need in cooking, and we believed *a priori* that they were subjected to a different set of choice criteria than the other varieties. The germplasm collectors helped to classify the varieties based on the name of the variety, the plant height according to the farmer, and whether the farmer considered the variety to be glutinous or not.[3]

Figure 6.1 presents the distribution of the number of varieties across ecosystems. The rainfed lowland and the upland ecosystems have almost the same total number of varieties, while the irrigated ecosystem has substantially fewer. There are other interesting differences among ecosystems. In the upland ecosystem, traditional varieties dominate, and modern varieties are few. In the irrigated ecosystem, only modern varieties are grown. In the rainfed lowland ecosystem, modern and traditional varieties coexist, although traditional varieties are more common than modern varieties. Glutinous varieties are present in both the upland and rainfed lowland ecosystems but are absent in the irrigated ecosystem. The distribution and importance of modern and traditional varieties across ecosystems are consistent with the expected pattern.

The genetic diversity of the varieties grown by farmers is also greater in the upland and rainfed ecosystems than it is in the irrigated ecosystem, as indicated by the Nei's unbiased expected heterozygosity, which was 0.24, 0.21, and 0.15 for the upland, rainfed lowland, and irrigated ecosystems, respectively. More attention should be given to the gradient of diversity that is observed across ecosystems, rather than to the absolute

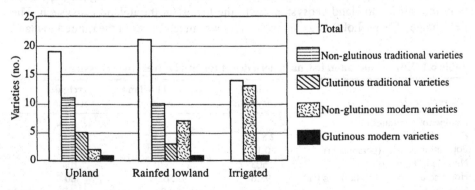

Figure 6.1. Number of Varieties by Type and Ecosystem.
Source: Bellon *et al.* (1998).

values, which depend on the type of genetic markers employed. A major qualitative difference exists between the diversity in the upland ecosystem and that of the irrigated and rainfed lowland ecosystems. In the upland ecosystem, varieties from both the indica and japonica[4] groups were grown, whereas indica varieties predominated in the other two ecosystems. Genetic diversity in rice in the upland ecosystem was not only higher but had a different genetic base.

6.4. FARMERS' PERCEPTIONS OF THEIR VARIETIES

To explain these patterns of genetic and varietal diversity in the rice varieties chosen by farmers, we examined the factors influencing their choice and the criteria they use. Informants were asked about certain quantitative characteristics of each variety they planted, such as its yield in good and bad seasons and its duration. Farmers were also asked to specify the advantages and disadvantages that they perceived in each of the varieties they planted.

Some questions were open-ended, and we interpret the frequencies associated with these positive and negative perceptions as reflecting the importance of these concerns to the farmers. These are expressed perceptions, not the results of controlled experiments. Lando and Mak (1994) observed that farmers' perceptions of grain quality in rainfed lowland rice differed notably from the results of laboratory evaluations. It is farmers' perceptions, however, rather than laboratory studies, that determine which varieties farmers will maintain or discard, and these perceptions therefore are the decisive factor for on-farm conservation.

Table 6.2 presents the average for the subjective yields reported by farmers for a good and a bad season, as well as for the duration of their varieties by type and ecosystem. Modern varieties were always superior to traditional ones for both a good

Table 6.2. Average of Variety Yields and Duration Reported by Farmers, by Rice Type and Ecosystem

	Upland ecosystem[a]			Rainfed lowland ecosystem				Irrigated ecosystem
	Non-glutinous MVs	Non-glutinous TVs	Glutinous TVs	Non-glutinous MVs	Non-glutinous TVs	Glutinous TVs	Glutinous MVs	Non-glutinous MVs
Yield (t/ha) in good season[b]	2.35	1.85	1.23	3.22	2.28	1.38	1.64	4.12
Yield (t/ha) in bad season[b]	0.98	0.80	0.41	2.56	1.68	1.21	1.64	1.59
Duration (days)[c]	118	129	133	125	179	142	120	114

Note: TVs = traditional varieties; MVs = modern varieties.
[a] Glutinous modern varieties not included since there was only one case.
[b] Subjective yield estimates.
[c] Days after sowing/transplanting to harvest. This number may overestimate actual duration because the farmers may leave the crop in the field for drying.

and a bad season. Glutinous varieties were inferior to non-glutinous ones in the rainfed ecosystem. Modern varieties were considered to be of shorter duration than traditional varieties, and hence in all ecosystems modern varieties were considered to be more productive per unit area per unit time (growing season) when compared with traditional ones. The puzzling question, then, is why have modern varieties not displaced traditional varieties in the rainfed lowland and upland ecosystems?

Table 6.3 shows the distribution of the positive traits reported by farmers in each ecosystem. Among these, traits related to consumption quality are the most important to farmers in both the upland and rainfed lowland ecosystems. High yield is the most frequently cited trait in the irrigated ecosystem, particularly if other components of yield are included, such as number of tillers, grain weight, and number. Even in the irrigated ecosystem, however, consumption traits are relevant. Furthermore, in the two rainfed ecosystems, farmers cited many more different traits related to consumption quality than in the irrigated ecosystem. Tolerances to abiotic and biotic stresses are of the next greatest importance for the upland and irrigated ecosystems, while these traits are cited infrequently by farmers in the rainfed ecosystem. Short maturation period was mentioned only by farmers in the irrigated and rainfed ecosystems. Management considerations, such as ease of harvesting and low use of inputs, were mainly cited by informants in the upland ecosystem.

Susceptibility to abiotic and biotic stresses dominates the disadvantages cited by farmers in all ecosystems (Table 6.4). The importance of specific stresses depends on the ecosystem. There seems to be a gradient in the frequency of traits mentioned by farmers from the upland to the irrigated ecosystems, increasing for susceptibility to abiotic stresses and decreasing for susceptibility to biotic stresses. Management considerations were repeatedly cited by farmers in the upland ecosystem, and to a lesser extent by those in the other two ecosystems. Late maturity was mentioned only by farmers in the rainfed lowland ecosystem.

Table 6.3. Distribution of Farmers' Responses about Positive Varietal Characteristics

Characteristic	Upland ecosystem (% farmers)	Rainfed lowland ecosystem (% farmers)	Irrigated ecosystem (% farmers)
High yield	9.3	6.7	24.2
Long panicle	2.3	1.0	–
More tillers	–	8.7	8.8
Good grain characteristics	1.2	4.8	11.0
Good milling quality	5.8	1.0	2.2
High price	–	7.7	2.2
Good consumption qualities (taste, texture, volume expansion, cooking)	54.7	56.7	16.5
Tolerance to abiotic stresses (drought, lodging)	7.0	4.8	15.4
Tolerance to biotic stresses (weeds, pests, diseases, birds)	12.8	1.0	12.1
Early maturity	–	4.8	6.6
Management (low inputs, easy to harvest)	7.0	2.9	1.1
Total	100	100	100

Source: Bellon *et al.* (1998).

The advantages cited by farmers are mainly associated with consumption quality and yield, while the disadvantages are associated with susceptibility to stresses and management requirements. However, the relative importance of traits differs by ecosystem. For example, although concerns about consumption quality are overwhelmingly important in both the upland and rainfed lowland ecosystems, their importance may not be related only to preferences in home consumption, but also to price premiums. Farmers in all ecosystems consume part of what they produce, but in the rainfed lowland ecosystem, informants observed that the most commonly grown traditional varieties (the so-called "Wagwag group" of varieties, which are considered to have the highest consumption quality) also fetch a higher price. To the extent that these ecosystems represent different levels of intensification, the data suggest that farmers' concerns and their relative importance change with intensification. Tolerance to wild pigs is a concern only in the upland ecosystem, while susceptibility to drought is important only in the rainfed ecosystems.

Table 6.5 presents evidence on the relative performance of modern and traditional varieties with respect to the traits identified above by ecosystem. The first column for each ecosystem is the relative frequency of the response from Table 6.3, and the remaining columns show the distribution of the response among the different types of varieties. For example, high yield in the rainfed lowland ecosystem had a relative frequency of 7% among all positive traits cited by farmers. Of those responses, 16% were associated with traditional non-glutinous varieties, while 83.3% were identified with modern, non-glutinous varieties.

In the upland and rainfed lowland ecosystems, high yield is associated mainly with modern varieties, while aroma and volume expansion are mainly linked with traditional varieties. In the rainfed lowland ecosystem, high price is associated with traditional varieties, while short duration is associated with modern varieties. Other consumption characteristics such as softness and good eating quality are primarily associated with traditional varieties in the upland ecosystem but are more evenly distributed among

Table 6.4. Distribution of Farmers' Responses about Negative Varietal Characteristics

Characteristic	Upland ecosystem (% farmers)	Rainfed lowland ecosystem (% farmers)	Irrigated ecosystem (% farmers)
Low yield	16.7	7.1	–
Poor grain characteristics	8.3	–	3.5
Poor tiller characteristics	–	3.6	3.9
Poor milling characteristics	–	–	3.5
Poor consumption characteristics	8.3	3.6	–
Susceptibility to abiotic stresses (drought, lodging)	8.3	35.7	44.8
Susceptibility to biotic stresses (pests, diseases, birds)	25	21.4	17.2
Late maturity	–	10.7	–
Management (high inputs, difficult to harvest)	33.3	17.9	17.2
Other (short stature, good only for dry season)	–	–	6.9
Total	100	100	100

Source: Bellon *et al.* (1998).

modern and traditional varieties in the rainfed lowland ecosystem. Resistance to drought is associated only with traditional varieties in the upland ecosystem, but more so with modern varieties in the rainfed lowland ecosystem.

Table 6.6 shows results regarding negative traits. Low yield was associated exclusively with traditional varieties in both ecosystems. Lack of drought resistance was associated with traditional varieties in the upland ecosystem and was associated more evenly between traditional and modern varieties in the rainfed lowland ecosystem. Long duration was a problem of traditional varieties in the rainfed lowland ecosystem. Susceptibility to biotic stresses was related to modern varieties in the upland ecosystem

Table 6.5. Frequency Distribution for Key Positive Traits, by Type of Variety and Ecosystem

Characteristic of variety	Upland ecosystem (% farmers)					Rainfed ecosystem (% farmers)				
	Total[a]	Non-glutinous TVs	Glutinous TVs	Non-glutinous MVs	Glutinous MVs	Total[a]	Non-glutinous TVs	Glutinous TVs	Non-glutinous MVs	Glutinous MVs
High yield	7.0	16.7	–	83.3	–	4.8	–	–	100	–
More tillers	–	–	–	–	–	8.7	55.6	–	44.4	–
High price	–	–	–	–	–	7.7	75	25	–	–
Good eating quality	7.0	66.7	16.7	16.7	–	21.2	50	4.5	45.5	–
Volume expansion	4.7	75	–	25	–	11.5	83.3	–	16.7	–
Aromatic	18.6	87.5	12.5	–	–	7.7	100	–	–	–
Soft	11.6	60	20	20	–	4.8	20	40	40	–
Very glutinous	3.5	–	66.7	–	33.3	7.7	–	75	–	25
Drought resistant	4.7	75	25	–	–	3.8	25	25	50	–
Short maturing	–	–	–	–	–	4.8	–	–	80	20

Note: TVs = traditional varieties; MVs = modern varieties.
[a] From Table 6.3 for responses of frequencies >4% for at least one type. Irrigated ecosystem excluded because only modern varieties are grown.

Table 6.6. Frequency Distribution for Key Negative Traits, by Type of Variety and Ecosystem

Characteristic of variety	Upland ecosystem (% farmers)					Rainfed ecosystem (% farmers)				
	Total[a]	Non-glutinous TVs	Glutinous TVs	Non-glutinous MVs	Glutinous MVs	Total[a]	Non-glutinous TVs	Glutinous TVs	Non-glutinous MVs	Glutinous MVs
Low yield	16.7	100	–	–	–	7.1	100	–	–	–
Fewer grains	8.3	–	–	100	–	–	–	–	–	–
Not drought resistant	8.3	100	–	–	–	25	28.6	14.3	57.1	–
Not lodging resistant	–	–	–	–	–	7.1	100	–	–	–
Late maturity	–	–	–	–	–	10.7	100	–	–	–
Susceptible to stem borer	–	–	–	–	–	10.7	100	–	–	–
Susceptible to pests	8.3	–	–	100	–	3.6	–	–	–	100
Requires high inputs	–	–	–	–	–	10.7	–	–	66.7	33.3
Shatters easily	16.7	100	–	–	–	7.1	50	–	50	–
Germinates easily	16.7	50	50	–	–	–	–	–	–	–

Note: TVs= traditional varieties; MVs = modern varieties.
[a] From Table 6.3 for responses of frequencies >4% for at least one type. Irrigated ecosystem excluded because only modern varieties are grown.

and with traditional and modern glutinous varieties in the rainfed lowland ecosystem. High use of inputs was a negative characteristic only of modern varieties in the rainfed lowland ecosystem.

These findings suggest that modern and traditional varieties complement each other, because they perform differently with respect to farmers' concerns. In both rainfed ecosystems, the lowland and upland, farmers seem to balance the modern varieties' higher productivity per unit area per unit of growing time against the traditional varieties' higher consumption quality and reliability. Glutinous varieties, which seem to be poorer in their overall performance, seem to be maintained by farmers because of these varieties' special consumption characteristics (they are used in special preparations such as desserts and candies). The balance among these concerns can change between ecosystems. The complete absence of traditional varieties from the irrigated ecosystem, for example, suggests that this balance can be broken by intensification.

6.5. FARMERS' PERCEPTIONS OF THE VARIETIES THEY HAVE DISCARDED

We asked farmers to name the varieties that they used to plant but have abandoned, as well as their reasons for doing so.[5] This list is subjective, reflecting changes that were important enough and recent enough for farmers to recall easily.

The most important reasons that farmers reported for discarding their varieties are presented in Table 6.7 by type of variety and ecosystem. These reasons encompass many of the categories that farmers cited as disadvantages of varieties they currently plant. In the upland ecosystem the main reasons for discarding traditional varieties were low yield, lodging, and unavailability of seed. Unique to this ecosystem were reasons such as a change of residence and abandonment of slash-and-burn agriculture. In the rainfed lowland ecosystem, the main reasons for discarding traditional varieties were long maturity, difficulty in harvesting and threshing, extension advice to discard the variety, and a reduction of yield after repeated planting (probably linked to disease). In contrast, farmers discarded modern varieties because of the desire to try new varieties, difficulty in managing several varieties, and susceptibility to drought, pests, or diseases. In the irrigated ecosystem, two reasons for discarding traditional varieties dominated farmers' responses: "late maturity that affects the next crop" and low yield. For the modern varieties, lack of seed, replacement by other modern varieties, and susceptibility to pests and diseases were cited frequently.

There are a few consistent patterns across ecosystems. In both the rainfed and irrigated ecosystems, susceptibility to pests and diseases, as well as the availability of new varieties and a desire to try them, are important reasons for discarding modern varieties. In the rainfed lowland ecosystem, where modern and traditional varieties coexist, the desire to try new varieties is a much more important reason to abandon modern varieties than traditional ones, suggesting that the modern varieties have closer substitutes. Long duration is an important explanation for abandoning traditional varieties in these two ecosystems, while low yield is particularly important in the irrigated ecosystem.

6.6. IMPLICATIONS FOR ON-FARM CONSERVATION

Traditional varieties are maintained and coexist with modern ones in the upland and rainfed ecosystems, but they have disappeared from the irrigated ecosystem. In the first two ecosystems, the favorable consumption quality of traditional varieties seems to balance their lower yield and longer duration. Higher yield and shorter duration are major advantages of modern varieties. In the irrigated ecosystem, this balance seems to be broken by the possibility of producing a second crop, which dramatically increases the opportunity cost of maintaining traditional varieties.

From the data in Table 6.2, it can be seen that the difference in average yields between traditional and modern non-glutinous varieties in the upland ecosystem is 0.5 t/ha (27%) for a good season and 0.18 t/ha (22.5%) for a bad season, while for the rainfed lowland ecosystem the difference in yields is 0.94 t/ha (41.2%) in a good season and 0.88 t/ha (72.7%) in a bad season. If the modern non-glutinous varieties are more competitive with respect to yield in the rainfed lowland compared to the upland ecosystem, it is not surprising that they are more widely planted.

Table 6.7. Farmers' Reasons for Discarding Varieties, by Type of Variety and Ecosystem

Reason	Upland ecosystem (% farmers)	Rainfed lowland ecosystem (% farmers)		Irrigated ecosystem (% farmers)	
	TVs	TVs	MVs	TVs	MVs
Low yield	10.8	6.9	6.3	20.8	8.2
Few tillers	–	6.9	–	–	6.6
Low market price	–	–	–	4.2	1.6
Not good to eat	–	3.5	–	4.2	1.6
Less resistant to drought	–	–	12.5	–	–
Lodges easily	8.1	–	–	–	3.3
Susceptible to pests/diseases	2.7	–	12.5	–	6.6
Susceptible to Tungro	2.7	–	–	4.2	11.5
Produces less after repeating planting	–	10.3	12.5	–	–
Liked by wild pigs	5.4	–	–	–	–
Late maturity affects next crop	–	–	–	29.2	1.6
Very long maturity	2.7	17.2	–	–	–
Non-shattering	–	–	–	8.3	1.6
Difficult to harvest with rake	5.4	13.8	–	–	–
Difficult to thresh	–	6.9	–	8.3	1.6
Difficult for widows to manage several varieties	–	3.5	18.8	–	–
Availability of new seed	–	–	–	8.3	–
Shift to other varieties	–	–	–	4.2	13.1
Want to try other varieties	–	6.9	25.0	–	–
No seed available	10.8	6.9	–	–	14.8
Changed residence	13.5	–	–	–	–
Stopped slash and burn	8.1	–	–	–	–
Changing climatic conditions	–	3.5	12.5	–	–
Discouraged by extension agent	–	10.3	–	–	–

Note: TVs = traditional varieties; MVs = modern varieties.

These figures provide an idea of the yield that farmers are willing to sacrifice for higher consumption quality, other characteristics and factors held constant. However, if we assume a similar average yield for the traditional non-glutinous varieties in the irrigated and rainfed ecosystems, and compare it to the yield of modern non-glutinous varieties, the difference is large for a good season (1.94 t/ha) and almost equal (−0.09 t/ha) for a bad season. In the irrigated ecosystem a second crop is possible with modern varieties but not with traditional varieties, because of their longer duration. Adding the yield of a second crop, the difference between growing one crop with a traditional variety is very large for a good season (6 t/ha) and remains important even in a bad season (1.5 t/ha). Clearly, the chance to grow a second crop as the ecosystem intensifies alters the opportunity cost, in terms of output forgone, of maintaining traditional varieties. Not surprisingly the key reason for abandoning traditional varieties in the irrigated ecosystem was "late maturity that affects the next crop" (Table 6.7).

This leads to two questions relevant for on-farm conservation of genetic resources. First, is there a means of providing incentives for maintaining traditional varieties as agriculture intensifies? Second, can the maintenance of traditional varieties enhance the genetic diversity present in an agroecosystem as agriculture intensifies?

Traditional varieties are not a homogeneous group. There are superior and inferior traditional varieties, both in terms of traits that farmers consider desirable and in terms of their contribution to genetic diversity. An example is the Wagwag group of traditional non-glutinous varieties, collected in the rainfed lowland ecosystem. These varieties account for most of the responses regarding positive traits of traditional varieties in that ecosystem[6] (Table 6.8). For example, out of all non-glutinous traditional varieties

Table 6.8. Characteristics of the Wagwag Varieties Relative to All Non-Glutinous Traditional Varieties in the Rainfed Ecosystem

	All non-glutinous traditional varieties	Wagwag varieties
Yield in good season (t/ha)	2.28	2.62
Yield in bad season (t/ha)	1.68	2.22
Duration (days after sowing)	178.5	175.8
Positive characteristics (% farmer responses)[a]		
More tillers	55.6	44.4
High price	75	75
Good eating quality	50	45.5
Volume expansion	83.3	58.3
Aromatic	100	75
Soft	20	–
Drought resistant	25	25
Negative characteristics (% farmer responses)[b]		
Low yield	100	100
Not drought resistant	28.6	28.6
Late maturity	100	100
Susceptible to stem borer	100	66.7
Shatters easily	50	50

[a] First column from Table 6.5.
[b] First column from Table 6.6.

deemed aromatic, the Wagwag varieties accounted for 75% of the answers, while for volume expansion, they accounted for 58.3% of favorable responses. Although they mature late, have low yields, and are susceptible to stem borers, their yields are relatively high in both good and bad seasons for non-glutinous traditional varieties.

Wagwag varieties also constitute a distinct genetic group within indica varieties, as shown by the isozyme analysis. Figure 6.2 presents a correspondence analysis of the isozyme data for all the varieties analyzed. The Wagwag group stands apart from the other varieties, in both the irrigated and rainfed lowland ecosystems. Although this specificity derives from a small number of alleles, the prominence of the Wagwag group is supported by data on polymorphism measured with both isozyme and microsatellite techniques (unpublished results, Genetic Resources Center, International Rice Research Institute).

Since the Wagwag group is genetically distinct from all other varieties grown in the irrigated ecosystem, the addition of the Wagwag group to the set of varieties grown by farmers in this ecosystem should increase genetic diversity among its rice cultivars. The varieties in this group, however, share the disadvantages of longer duration and lower yield than modern varieties. A breeding or management intervention that decreases the duration and increases the yield of Wagwag varieties might enhance their attractiveness to farmers while contributing to the genetic diversity of the ecosystem, assuming that their allele structure is maintained.

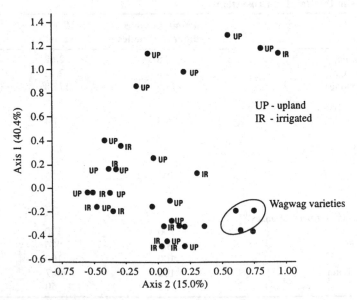

Figure 6.2. Correspondence Analysis of the Isozyme Data.
Source: Bellon *et al.* (1998).

6.7. CONCLUSIONS

By relating farmers' perceptions of varieties' characteristics to the varietal and genetic diversity that farmers maintain across rice ecosystems, our approach recognizes that certain varieties can make an important genetic contribution to an agricultural system, while serving the needs of farmers. Changes in opportunity costs of growing traditional varieties occur as agriculture intensifies, and in this study, they are mainly associated with trade-offs between yield and consumption quality.

The challenge for on-farm conservation is to decrease these opportunity costs. Meeting this challenge requires an understanding of farmers' perspectives, in order to identify the nature of the trade-offs and the opportunity costs involved. It also requires analysis of the genetic diversity found in the varieties they grow. Research on farmers' incentives to cultivate certain varieties and the relationship of these varieties to genetic diversity offers a richer approach to developing on-farm conservation initiatives than the blanket recommendation that farmers continue to cultivate traditional varieties.

Notes

1 When this research was conducted, M. Bellon was with the Genetic Resources Center, International Rice Research Institute, Philippines.

2 In this chapter, "modern variety" refers to varieties developed and released by the formal plant breeding system after 1960. Usually they have short stems, are photoperiod insensitive, and have a good response to fertilizer. "Traditional variety" refers to indigenous varieties and to introduced varieties produced by the formal plant breeding system before 1960. These are usually tall and many of them are sensitive to photoperiod.

3 If the name included a number, a variety was considered to be modern. In one case, a variety was classified as modern, even though its name included no number, because the farmer declared that the variety was short. A variety was considered to be glutinous if its name included the word *diket* ("glutinous").

4 Asian rice (*Oryza sativa*) is differentiated into two major ecogeographical races as a result of isolation and selection: 1) indica, which is adapted to the tropics, and 2) japonica, which is adapted to temperate regions and tropical uplands (IRRI, 1993b).

5 No distinction was made between glutinous and non-glutinous varieties, because farmers only provided names, and the collector could not examine a sample of the variety.

6 There is an inconsistency in responses about drought resistance, because some farmers considered varieties to be resistant while others disagreed. The level of resistance displayed by a variety is relative. Since farmers plant each variety as part of a set of varieties, and sets differ among farmers, variation in opinions may reflect the fact that each farmer had a unique frame of reference.

References

Balakrishna, P. 1996. Role of traditional cultivars and *in situ* conservation in sustainable agriculture: A case study in rice (*Oryza sativa* L.). *Plant Genetic Resources Newsletter* 107: 45–49.

Bellon, M. R., J. L. Pham, and M. T. Jackson. 1997. Genetic conservation: A role for rice farmers. In N. Maxted, B.V. Ford-Lloyd and J.G. Hawkes (eds.), *Plant Conservation: The In Situ Approach*. London: Chapman and Hall.

Bellon, M. R., J. L. Pham, L. S. Sebastian, S. R. Francisco, G. C. Loresto, D. Erasga, P. Sanchez, M. Calibo, G. Abrigo, and S. Quilloy. 1998. Farmers' knowledge on varietal selection in three rice ecosystems in the Philippines: Implications for on-farm conservation. Mimeo. Manila, Philippines: International Rice Research Institute (IRRI).

Byerlee, D. 1994. *Modern Varieties, Productivity, and Sustainability: Recent Experience and Emerging Challenges*. Mexico, D.F.: International Maize and Wheat Improvement Center (CIMMYT).

Chang, T. T. 1994. The biodiversity crisis in Asia crop production and remedial measures. In C. I. Peng and C. H. Chou (eds.), *Biodiversity and Terrestrial Ecosystems*. Institute of Botany, Academia Sinica Monograph Series No. 14. Taipei: Academia Sinica.

David, C. C., and K. Otsuka. 1994. *Modern Rice Technologies and Income Distribution in Asia.* Boulder: Lynne Rienner and the International Rice Research Institute (IRRI).

IRRI (International Rice Research Institute). 1992. *Challenges and Opportunities in a Less Favorable Ecosystem: Rainfed Lowland Rice.* IRRI Information Series No. 1. Manila: IRRI.

IRRI (International Rice Research Institute). 1993a. *Challenges and Opportunities in a Less Favorable Ecosystem: Upland Rice.* IRRI Information Series No. 2. Manila: IRRI.

IRRI (International Rice Research Institute). 1993b. *IRRI Rice Almanac 1993–1995.* Manila: IRRI.

IRRI (International Rice Research Institute). 1994. *Challenges and Opportunities in a Less Favorable Ecosystem: Flood-prone Rice.* IRRI Information Series No. 4. Manila: IRRI.

Khush, G. S. 1984. *Terminology for Rice-growing Environments.* Manila: International Rice Research Institute (IRRI).

Lando, R. P., and S. Mak. 1994. *Cambodian Farmers' Decisionmaking in the Choice of Traditional Rainfed Lowland Rice Varieties.* IRRI Research Paper Series 154. Manila: International Rice Research Institute (IRRI).

Vaughan, D. A., and T.T. Chang. 1992. *In situ* conservation of rice genetic resources. *Economic Botany* 46: 368–383.

7 AGRONOMIC AND ECONOMIC COMPETITIVENESS OF MAIZE LANDRACES AND *IN SITU* CONSERVATION IN MEXICO

H. Perales R., S. B. Brush, and C. O. Qualset

7.1. INTRODUCTION

When public attention initially focused on the conservation of crop genetic resources in the late 1960s, scientists advocated *ex situ* conservation because of their belief that modern varieties would inevitably replace traditional varieties (Frankel, 1970a, 1970b). This belief was based on two key assumptions. The first, explicit assumption is that modern varieties are always superior to traditional varieties in yield and economic profitability. A second, implicit assumption is that all farmers share the objective of maximizing expected profits. Since superior modern varieties would sooner or later replace landraces, *in situ* conservation of crops was dismissed *a priori* as non-viable. Interventions such as subsidies would be required to maintain cultivation of traditional varieties in the face of the high opportunity costs of growing them (Ford-Lloyd and Jackson, 1986), especially as greater food production was needed to feed the ever-increasing world population.

Since then, researchers have demonstrated that, on the contrary, *in situ* conservation of crops occurs *de facto* in many places, particularly in centers of crop origin and diversification (Bellon and Brush, 1994; Brush, 1995; Dennis, 1987; chapters 6 and 8). Post-harvest processing and storage losses, or inferior consumption quality of modern varieties, may outweigh their apparent yield advantage over traditional varieties (Byerlee, 1996), especially for farmers who consume a part of their harvest (Smale and Heisey, 1998). Bellon (1996) has proposed that small-scale farmers' choice to grow more than one variety simultaneously reflects their need to address numerous concerns, which no single variety is likely to satisfy. Brush (1989) included cultural factors, such as an emphasis on farm diversity and on culinary and other cultural preferences.

From the conventional perspective of the microeconomics of technology adoption in agriculture, *de facto* conservation of traditional varieties by farmers may represent their partial adoption of modern varieties. Growing both traditional and modern varieties can be understood as the outcome of seeking to satisfy objectives other than profit maximization, such as the maximization of utility in a household decisionmaking model. Microeconomic explanations for partial adoption have been reviewed and tested by Meng (1997) and Smale, Just, and Leathers (1994). These include factors related to the demand for and supply of modern varieties, including risk and uncertainty, missing markets, and differential soil quality (Bellon and Taylor, 1993).

Some researchers have based their explanations on determinants of the supply of modern varieties. Frankel, Brown, and Burdon (1995) stressed that a main factor in the conservation of traditional varieties could be the lack of local plant breeders to produce varieties "tuned with the agroecosystem" (see also chapter 15). Brush (1989) proposed a similar idea with respect to higher research costs for adapting modern varieties to heterogeneous farming systems, and Byerlee (1996) observed that for some regions the international research system has been unable to produce varieties possessing a yield advantage. In particular communities, landraces may be competitive or superior in yield under local conditions to the modern varieties that are supplied to farmers. If so, growing traditional varieties instead of modern varieties might express the profit-maximizing behavior of farmers.

The purpose of this chapter is to examine why maize landraces are maintained in the Amecameca and Cuautla Valleys of Mexico. Survey evidence demonstrates that farmers in these areas cultivate only a few traditional varieties, which occupy most of the maize area. These traditional varieties appear to be competitive with modern varieties in terms of yield and net income from production. In this region, farmers whose sole objective is to maximize profits would be indifferent to the choice between their own traditional varieties and the modern varieties available to them. Either the formal (public or private) seed system has been unable to produce modern varieties with yield and/or income advantages over the traditional varieties grown by these farmers, or farmers themselves are capable of effectively adapting the genetic pools provided by their traditional materials to the natural and social environment.

What then determines the choice among modern varieties and major traditional varieties for these farmers? The data also indicate that maize as a crop generates nil or negative profit, which is consistent with the notion that farmers are pursuing objectives other than profit maximization. Several minor varieties are cultivated by a few farmers on small areas, and the reasons for maintaining them appear to include their special characteristics and farmers' interest in experimentation.

7.2. DATA SOURCES AND DEFINITIONS

Following exploratory visits and informal interviews in the Amecameca and Cuautla Valleys, four communities were selected for study. The communities differed in the major maize types and other main crops that farmers cultivated. Within each community, 25% of households were selected for sampling based on a street map obtained from local authorities and subsequently verified and adjusted by inspection. The survey was implemented with 13–18% of the households, depending on the community. The sample included both farming and non-farming households. From each survey household, researchers requested a sample of 33 seed ears of each type of maize grown during the preceding season. Information regarding the local names of varieties, the source of seed, and the number of years since seed was originally obtained was recorded. Details about seed management, farmers' soil taxonomies, production activities for each crop and plot of land, family characteristics, and sources of income were noted. Questions about production and income referred to the preceding season, while questions about maize seed management included previous years. Details of the survey data and related calculations are available in Perales (1998).

Cost and income were calculated based on survey data for each farmer and are presented on a per-hectare basis. Variants of maize types that farmers had given separate names were aggregated when it made sense to do so based on shared morphological characteristics and growing environment. For example, in the Tlaltetelco community, the maize type known as Ancho was combined with Ancho-pozolero, and in the Ayapango community, Crema, Blanco, and Criollo were combined. Net economic returns from maize varieties and other crops were calculated in two ways. The more favorable estimate of net income includes all income from production sold as well as the value of the production consumed and the government subsidy, but only cash costs of production. For the less favorable estimate, gross income was also calculated as the total value of production and subsidy, but total costs included cash outlays as well as non-monetary costs assessed at local market prices for family labor, land, and traction. Data on net returns were converted to US dollars based on the 1995 average exchange rate (Mx\$ 7.0/US\$ 1.0).

In one of the survey communities, Tlaltetelco, a trial was established to compare the most widely grown local landrace, Ancho-pozolero, with a hybrid (Pioneer 3288). Pioneer 3288 had been recommended for that community by Mexico's national agricultural research institute, INIFAP (Instituto Nacional de Investigaciones Forestales, Agrícolas y Pecuarias). The purpose of the trial was to compare the local variety with a technology that was expected to be superior. The trial was designed and supervised by researchers but managed by farmers, and included plant density and fertilizer treatments with replicates in large plots (64–100 m^2). Details of trial design and results are found in Perales (1998).

In this chapter, the term "modern varieties" refers to the products of the formal plant breeding system. "Advanced generations" of modern varieties are those that were bought from commercial sources at least one year before the survey season (1995) and have been cultivated and selected by farmers, including both hybrids and improved open-pollinated materials. "Traditional varieties" are defined as all other varieties— the continually changing, open-pollinated maize populations that farmers manage from season to season. The category of "special cases" refers to varieties that may derive from modern varieties and purposeful mixtures of hybrids and traditional varieties. Race names follow Wellhausen, Roberts, and Hernández X. (1951).[1]

7.3. SITE DESCRIPTION

The communities selected for the study form part of two distinct agroecological regions (Figure 7.1) that are separated in elevation by about 2,000 m. The community of Ayapango, located in the southeastern part of the Basin of Mexico in the Central Highlands, is less than 40 km from Mexico City, near Amecameca in the state of Mexico. The villages in the state of Morelos are at the upper east side of the Balsas Depression watershed in the Cuautla Valley. Tlaltetelco is located in *los altos* (the highlands) of Morelos, while Tecajec and López Mateos are in the *tierra caliente* (literally, "hot land") of Morelos. Part of the land endowment of López Mateos has irrigation, which is used for vegetables, but all other crops in the four communities, including maize, are managed as rainfed crops. Selected characteristics of the study communities and their households are presented in Table 7.1.

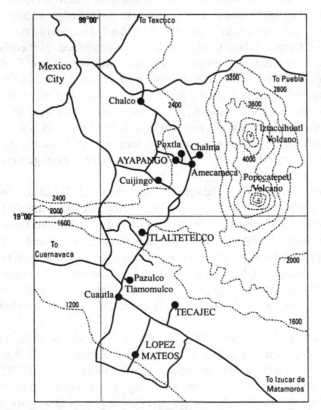

Figure 7.1. Communities Studied and Reference Points in the Region.

Ayapango has a cool, humid climate. Frosts are common between November and February, but summer conditions permit a growing season of more than six months. Rainfall is substantial during the crop season, starting in May and ending by October, with a small dry period during August or September. Winter rainfall at Ayapango is sufficient to start the maize crop before the rainy season begins. Soils are deep and friable, of volcanic origin, with good drainage. The climate becomes warmer and drier as the elevation decreases sharply towards Morelos in the south. Tlaltetelco has warmer weather than the highlands, but nights are cool and it is not as hot as in Tecajec and López Mateos. In the communities of Morelos rainfall patterns are the same as in Ayapango. In Tlaltetelco, a rainfed tomato crop is possible. In Tecajec and López Mateos sorghum is an important crop because of the drier conditions. Soils are well drained and good for agriculture, although soil quality is better in Ayapango than in Tlaltetelco, Tecajec, and López Mateos.

More than two-thirds of the land in these communities is *ejido* land. The *ejido* system is a form of collective land tenure, currently under reform to privatize rights. Over one-half of the farmers at Ayapango hold some land as private property, while only one-fourth at Tlaltetelco and Tecajec and none at López Mateos farm privately held fields. About one-tenth of the land in the four communities is rented. Only one-half of the households in Ayapango obtain their main income from their own farms. The percentage relying on farm income is between 60% and 80% for the communities

Table 7.1. Selected Characteristics of Study Communities and Households, Amecameca and Cuautla Valleys, Mexico, 1995

Characteristic	Ayapango	Tlaltetelco	Tecajec	López Mateos
Community				
State	Mexico	Morelos	Morelos	Morelos
Municipality	Ayapango	Atlatlahucan	Yecapixtla	Tepalcingo
Population (1990)	1,943	8,91	1,044	428
Elevation (masl)	2,400	1,700	1,400	1,200
Mean temperature (°C)	14.8	20.2	23.0	24.8
Mean precipitation (mm)	1,050	1,020	950	850
Other	–	–	–	Irrigation
Percent area planted to maize	93	74	65	32
Principal other crops	Tomato	Sorghum	Sorghum	Sorghum, onion
Household				
Total number of households	355	210	159	112
Number of survey households	46	32	21	20
Average age of household head for whom main income is own farm (yr)	60.1	47.9	48.8	48.6
Percent households for which main source of income is own farm	52	59	81	70
Other income	Salary	Salary	Salary	Remittances
Mean ha maize/farm	2.2	1.9	1.9	0.9
Mean number varieties/farm	1.57	1.76	1.50	1.42

Source: Mean annual temperature and mean precipitation data are from nearest weather stations (García, 1973). Population data from INEGI (1994a, 1994b). Remainder of data from Perales (1998).

of Morelos. Salary income, commonly derived from agricultural labor, is important for households in Ayapango, Tlaltetelco, and Tecajec but almost negligible in López Mateos. Migration (to the United States) is significant only for López Mateos. The average age of the head of the farm household is almost 50 years for the communities of Morelos and 60 years for Ayapango; in Ayapango, the average age of heads of non-farm households is only 40. Most of the attrition from farming into non-farm occupations has occurred during the last ten years.

In Ayapango all farmers cultivate maize, and maize occupies more than 90% of the area under cultivation. Farmers also produce small quantities of oats, beans, squash, and broad beans, and the latter three crops are intercropped with maize. At Tlaltetelco 90% of the farmers cultivate maize on three-fourths of the total cultivated area, and 60% cultivate tomato on one-fourth of the area. Very small quantities of other crops are present. Beans are intercropped with tomato, and squash with maize. At Tecajec, all farmers cultivate maize on two-thirds of the crop area, and 60% of the farmers cultivate sorghum on one-third of the area; very small quantities of other crops are present, with beans and squash intercropped with maize. At López Mateos, 93% of the farmers grow maize on one-third of the cultivated area, one-half of the farmers cultivate sorghum on almost one-half of the area, and one-half of the farmers cultivate irrigated onions or zucchini squash on 20% of the area (Perales, 1998).

7.4. RESULTS OF THE SURVEY AND TRIAL

7.4.1. Varietal Diversity
Maize types with white grain occupied more than 85% of the area planted to maize across all of the study communities (Table 7.2). Varieties with blue, red, and yellow grain, and other minor varieties, were grown by about one-fourth of the survey farmers. Each of these varieties typically covered less than 5% of the maize area.

In the highlands (2,400 masl), white and colored varieties are variants of the Chalqueño race, which together were grown on more than 95% of the area planted to maize. The only other race present in the highlands is Cacahuacintle, which was cultivated by a few farmers on small areas. Farmers grew no modern varieties, with the exception of a backyard type which had reputedly originated from a hybrid over 20 years ago. Only in this community was a blue type cultivated on almost 10% of the area planted to maize.

At the intermediate elevation in Tlaltetelco (1,700 masl), two types of maize dominate. The traditional variety, locally known as Ancho-pozolero or Pozolero, is related to an historic local type, known only as Ancho, which is still present in very small quantities. Farmers recognize Ancho-pozolero as distinct from Ancho although both are classified in the Ancho race. Ancho-pozolero commands about twice the price of all other types of maize, because of a specialty dish prepared from it, and it is cultivated on almost two-thirds of the maize area in this community. The second major maize type is referred to as "hybrid," which farmers agree was introduced into the area (possibly from Celaya) by informal seed sources or farmers themselves over 20 years ago.[2] This variety is cultivated on almost one-third of the maize area. Blue and red types of the Ancho race are also found and are cultivated on very small areas.

In the lowlands of Morelos (Tecajec and López Mateos), commercial hybrids are sown, all of them white, along with advanced generations of hybrids; a local, traditional variety, Delgado; and an introduced, traditional variety, Tehuacan. Some farmers at Tecajec mixed true or advanced-generation hybrids purposely with the seed of traditional varieties (see "special case" in Table 7.2). An historic traditional variety known as Pepitilla or Delgado (also of the race Pepitilla) seems to be disappearing. The area under the four major types is relatively evenly partitioned, with the exception of Delgado, which has a smaller share. In these two communities, blue and red maize similar to Delgado are found, and several other white types were cultivated on small areas.

The average number of maize types per farmer was less than two (Table 7.1), and the maximum was four types per farmer. One-half or more of the farmers in the study sites cultivated only one of the major varieties. When farmers had two or more types of maize, the most common pattern was a single white type of maize, combined with a blue maize or other colored maize. At Tlaltetelco one-third of the farmers planted the two major types, and in Tecajec and López Mateos one-fifth of the farmers had both a landrace and a hybrid or an advanced generation of a modern variety.

Table 7.2. Percentage of Farmers and Area Cultivated by Maize Type, Amecameca and Cuautla Valleys, Mexico, 1995

Maize type	Ayapango (2,400 masl)		Tlaltetelco (1,700 masl)		Tecajec (1,400 masl)		López Mateos (1,200 masl)	
	% farmers	% maize area	% farmers	% maize area	% farmers	% maize area	% farmers	% maize area
Traditional varieties								
White varieties								
White	96	87	–	–	–	–	–	–
Ancho	–	–	14	10	–	–	–	–
Ancho-pozolero	–	–	71	57	5	0.5	–	–
Delgado	–	–	–	–	25	14	7	1
Tehuacan	–	–	–	–	10	5	43	28
Cacahuacintle	11	1	–	–	–	–	–	–
Other	–	–	–	–	5	2	–	–
Colored varieties								
Blue-black	32	9	24	5	15	3	7	1
Red	7	0.2	10	0.5	5	0.3	7	2
Yellow	7	3	–	–	–	–	–	–
Modern varieties								
Hybrids	–	–	–	–	25	36	43	34
Advanced generation	–	–	–	–	65	31	31	35
Special cases								
"Hybrid" landrace	–	–	52	30	–	–	–	–
Hybrid x landrace	–	–	–	–	10	9	–	–
Total (farmers and ha)	28	59	21	40	20	39	14	13

Note: Based on farmers' named varieties. White and colored traditional varieties in different communities are not the same maize type even when they share the same name. Hybrids include several types. Advanced generations of modern varieties include hybrids and other modern types.

The dominance of a few major maize types in these communities holds for the other published studies of maize in Mexico. For example, Bellon and Brush (1994) found that three types of maize covered about 80% of the area under maize in the community of Vicente Guerrero in Chiapas. Louette, Charrier, and Berthaud (1997) found that in the Cuzalapa Valley of Jalisco three varieties were grown on more than 70% of the maize area. In both of these studies, the number of maize types planted to large areas was small compared to the total number of varieties. These findings suggest that farmers in Mexican communities maintain only a few varieties on a large scale.

Farmers' explanations indicate that the continued cultivation of minor varieties reflects special characteristics and qualities that are specific to individual farmers or communities, rather than their yield or economic competitiveness with other major varieties. For example, in the study communities, blue and red varieties are cultivated because they are sweeter when consumed in particular dishes. Some farmers state that they keep a variety because their father gave it to them. Experimentation also plays a role in the cultivation of minor varieties. In the communities studied here, more than one-half of the farmers had tried at least one different variety in the past five years, and in the community where the grain yields were lowest more minor types were cultivated. In the Louette, Charrier, and Berthaud study, many of the minor varieties were tested and abandoned within a few cycles because they proved inferior in one respect or another to varieties already grown. Farmer experimentation with new varieties is common worldwide (Johnson, 1972; Richards, 1986; Dennis, 1987; Feder, Just, and Zilberman, 1985).

7.4.2. Yields of Major Maize Types

Farmers' yield estimates are summarized in Table 7.3. Within communities, mean yields are similar across major maize types. On average, hybrids do not appear to yield better than the traditional varieties in Tecajec and López Mateos (non-significant by unpaired t-test). Yields are generally low in Tecajec. In Tlaltetelco the yield of the hybrid is larger than that of the Ancho-pozolero (statistically significant at 0.1 by unpaired t-test with unequal variances). At Ayapango, as mentioned above, there is only one major white maize material.

No statistical difference was found for yield in two out of three trials between the traditional maize and the Pioneer 3288 hybrid. In the trial where the yield difference was statistically significant, Ancho-pozolero had the higher yield. Yields of Ancho-pozolero were similar to those of the hybrid under high fertilization levels and high density levels. The highest overall means among treatments were similar for Ancho-pozolero and the hybrid, at 5.2 and 5.1 t/ha, respectively.

It could be argued that in the study communities the profitability of modern varieties is inferior to that of the major traditional maize varieties because the environmental or production conditions are marginal. Two findings contradict this argument. First, traditional varieties dominate maize area completely in Apayango, where environmental conditions are favorable. Soils are deep and fertile, moisture is good, and the growing season is long. Although this environment represents the adaptive limit for maize production, maize has been grown there for several thousands of years (Sanders, Parsons, and Stanley, 1979). Second, the trial comparing Ancho-pozolero to the recommended maize hybrid demonstrates that at least Ancho-pozolero has ample potential to respond to increasing improvement of production conditions. The range of plant densities (35,000–55,000 plants/ha) and fertilizer levels (60:46 to 180:138 $N:P_2O_5$) included in the experiment cannot be viewed as marginal for production from the agronomic point of view.

Perhaps the expectation that modern varieties should yield better than traditional varieties has resulted from trials conducted on experiment stations, where only one or few traditional varieties were chosen as a basis of comparison. Nevertheless, other data from the highlands of Mexico also suggest that modern varieties have not performed better than traditional varieties (CIMMYT, 1969, 1973; Muñoz, Carballo, and González, 1970; Cervantes and Mejia, 1984). In 17 on-farm trials conducted in Veracruz, Mexico, a modern variety produced 0.4 t/ha more than a traditional variety (19% higher) under either improved or farmers' management (Perrin, 1975). In a second set of ten trials, the yield of the traditional variety exceeded that of a modern variety by 0.4 t/ha without fertilization (16% higher), and with fertilization the yields were equivalent (Perrin, 1976). In other crops, the retention of traditional varieties may also reflect superior yields (Dennis, 1987; Zanatta *et al.*,1996).

7.4.3. Economic Returns from Production of Maize and Other Crops
In the study communities, farmers do not recognize differences in crop management practices or the use of purchased inputs for growing traditional and modern varieties. The inputs they use and the management practices they follow depend on their cash situation at the beginning of the season, rather than on maize type. For a given amount

Table 7.3. Yields Reported by Survey Farmers by Maize Type, Amecameca and Cuautla Valleys, Mexico, 1995

Maize type	Ayapango (2,400 masl)		Tlaltetelco (1,700 masl)		Tecajec (1,400 masl)		López Mateos (1,200 masl)	
	Mean yield (t/ha)	Standard error	Mean yield (t/ha)	Standard error	Mean yield (t/ha)	Standard error	Mean yield (t/ha)	Standard error
Traditional varieties								
White	1.9	0.13	–	–	–	–	–	–
Ancho-pozolero	–	–	1.5	0.17	–	–	–	–
Delgado	–	–	–	–	1.3	0.40	–	–
Tehuacan	–	–	–	–	–	–	2.7	0.45
Modern varieties								
Hybrids	–	–	–	–	0.9	0.27	2.3	0.42
Advanced generation	–	–	–	–	1.2	0.13	1.7	0.35
Special case								
"Hybrid" landrace	–	–	2.3	0.39	–	–	–	–
All types	1.8	0.12	1.8	0.17	1.2	0.11	2.3	0.23

Note: Based on farmers' named varieties. White traditional varieties in different communities are not the same maize type even when they share the same name. Hybrids include several types. Advanced generations of modern varieties include hybrids and other types.

of cash, inputs and treatments are allocated proportionately to the seed planted, across the major maize types a farmer cultivates. Minor maize types may receive fewer inputs or less management, but this is more a result of their status as a minor economic activity than it is a recognition of any distinct agronomic requirements they may have. Even though the quantity of different inputs and of the relative amount of monetary and non-monetary costs varies between communities, the mean total cost per hectare ranged only between US$ 380 and US$ 450 for the communities, with an overall average of US$ 410 (Perales, 1998).

Maize output was also valued similarly for hybrids and traditional varieties. Hybrids were not more likely to be sold than traditional varieties. A larger proportion of maize production was sold in the communities of Ayapango and Tlaltetelco, where only traditional varieties were grown, than in Tecajec and López Mateos, where modern varieties were also grown. Sales of maize may be related more to the productivity of the environments or periodic needs for cash than to the type of maize grown. Nor was value in consumption sacrificed with hybrids; farmers viewed modern varieties and traditional varieties as producing *tortillas* of similar quality. Finally, farmers also reported little difference in storage quality between traditional and modern varieties.

Table 7.4. Mean High and Low Estimates of Net Income per Hectare (US$) of Major Maize Types for Amecameca and Cuautla Valleys, Mexico, 1995

Maize type or crop	Ayapango (2,400 masl)		Tlaltetelco (1,700 masl)		Tecajec (1,400 masl)		López Mateos (1,200 masl)	
	High	Low	High	Low	High	Low	High	Low
Traditional varieties								
White	190	10	–	–	–	–	–	–
Ancho-pozolero	–	–	90	–130	–	–	–	–
Delgado	–	–	–	–	30	–220	–	–
Tehuacan	–	–	–	–	–	–	240	–10
Modern varieties								
Hybrids	–	–	–	–	–70	–330	230	10
Advanced generations of modern varieties	–	–	–	–	30	–190	–70	–270
Special case								
Advanced generation hybrid	–	–	110	–100	–	–	–	–
All types of maize	170	0	100	–100	30	–200	160	–60
Tomato	–	–	740	300	–	–	–	–
Sorghum	–	–	–	–	230	130	270	190
Onion	–	–	–	–	–	–	1,160	840

Note: High estimate takes into account all income (cash and consumption) but only cash expenditures.
Low estimate considers all income (cash and consumption) and all expenditures (cash and opportunity cost for family labor, land, traction). White traditional variety in different communities is not the same type. The average for all types is a weighted mean; includes colored and minor varieties.

Given similar costs of production, similar prices, and similar yields, few or no differences were found within a community for either measure of net income from the production of major maize types (Table 7.4). Table 7.5 presents the partial budget for communities where hybrids are planted. The cost of hybrid seed represents 15% of the monetary expenses in both cases, but only 7% if the opportunity costs of non-monetary inputs are considered. The two estimates for net income are similar for the traditional variety (Delgado or Tehuacan) and the hybrids within the community, but these two types are different from the advanced-generation hybrid for López Mateos (statistically significant at 0.1 level by unpaired t-test). In Tlaltetelco the lower yield of Ancho-pozolero in relation to the hybrid was compensated by a higher price (MX$ 1.6/kg compared with MX$ 0.8/kg in 1995), resulting in comparable estimates of net income. Negative estimates of net income were common, and positive estimates were small in magnitude. Roughly 68% of high estimates of net income were less than US$ 100/ha, while 87% were below zero for the low estimates (Figure 7.2). It is important to note that farmers described the cropping season in the survey year as "normal."

The other crops produced by survey farmers generated higher net incomes than maize (Table 7.4). In tomato production, 3 in 16 farmers had a negative net income for the high estimate and 5 had a negative net income for the low estimate. In onion

Table 7.5. Partial Budgets (US$/ha) for Major Maize Types in Study Communities where Hybrids are Cultivated, Morelos, Mexico, 1995

Cost category	Tecajec			López Mateos		
	Traditional	Hybrid	Advanced-generation hybrid	Traditional	Hybrid	Advanced-generation hybrid
Monetary costs	200	200	170	170	200	320
Non-monetary costs	240	260	210	240	210	200
Seed	0	30	0	0	30	0
Total cost	440	490	390	410	440	520
Income sale	30	20	30	110	240	50
Income consumption	180	140	170	240	170	220
Total gross income	220	160	200	410	460	260
High estimate net income						
Mean	30	–70	30	240	230	–70
Standard error	20	50	50	60	70	130
Low estimate net income						
Mean	–220	–330	–190	–10	10	–270
Standard error	80	100	60	50	90	100
n	8	6	12	6	6	5

Note: Total gross income includes subsidy when received by farmer. High and low estimate of net income as in Table 7.4. Cost of own seed less than US$ 5.

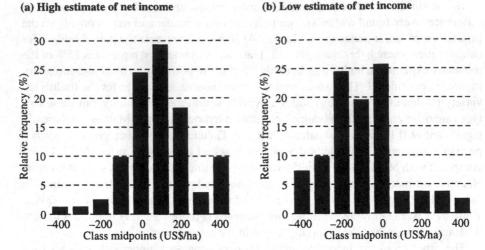

Figure 7.2. Distribution of the High (a) and Low (b) Estimate of Net Income (US$) per Hectare of Maize and per Farmer.
Note: n = 82.

production, 1 case in 7—a very small-scale producer—generated a negative net income for onions. Only 1 case in 20 had a negative net income for sorghum for the high estimate.

7.5. WHY DO FARMERS GROW MORE THAN ONE MAJOR MAIZE VARIETY?

The data demonstrate that no major maize type, including the maize hybrids, has a relative advantage in either yield or net income per hectare. In other words, the varieties are substitutes with respect to yield and/or expected economic returns. These results imply that if farmers grow more than a single major variety, they must be considering other factors. Other factors include characteristics of the variety that are not reflected in expected yield, various forms of risk, differential performance by soil type or environment on farms, and/or missing markets for related inputs or outputs.

All farmers in the four communities agreed that yield was an important consideration in the choice of their maize types, but they also mentioned characteristics other than yield as important. In all communities there was consensus that blue and red types of maize were sweeter and thus better suited for particular dishes. In all study communities except Ayapango, the market for colored maize types is limited. At Apayango, the blue maize type is now more important than it was some 20 years ago, possibly because a market developed for weekend tourism from Mexico City. Nevertheless, this emerging market has not resulted in the displacement of the dominant white type.

Differential performance of varieties by soil type or environmental conditions, as in the case presented by Bellon and Taylor (1993) for Vicente Guerrero, Chiapas, is interpreted here in terms of agronomic competitiveness. For the communities studied in the valleys of Amecameca and Cuautla, no systematic relation was found between the soil types recognized by farmers and the use of a particular variety.

The nature of market demand for grain matters, since it is rare to find truly autarkic farmers. The yellow type of maize produced in Ayapango is said to yield well, but because of its weak market, few farmers cultivate it. At Tlaltetelco the Ancho-pozolero seems to yield less than the hybrid landrace, but a price premium that reflects a socially recognized quality differential compensates for the lower yields.

Maize grain markets seem to be functioning for all the communities studied. Farmers in Ayapango, which is closest to the Mexico City market, do not grow modern varieties, while in the more isolated communities of Tecajec and López Mateos, hybrids have been adopted. Many farmers in Tecajec and López Mateos are short of cash, and for all practical purposes, there is no credit market in these communities. This suggests that those who cultivate advanced generations of hybrids are cash-poor, small-scale farmers.

Risk is an explanation that cannot be ruled out with the present study, since a full analysis of production and economic risk has not been conducted. The application of portfolio theory to farmers' choice of variety implies that if there are two alternative varieties, in which higher mean yield or income can be traded for lower variance in yield or income, the optimal solution depends on the risk attitudes of the farmer and may include the choice to grow both (Feder, 1980). Other criteria for decisionmaking under risk, such as minimizing the risk of obtaining a yield or net income in the "disastrous" tail of the distribution, might also be relevant. In general, however, other income sources among many of the farm households in the study communities are likely to insure at least in part against yield risk and the risk of negative net income from maize production (Table 7.1).

7.6. WHY DO FARMERS GROW MAIZE?

In the study communities, the opportunity cost of cultivating a traditional variety of maize is not the earnings foregone from the cultivation of modern varieties. Instead, the opportunity cost of growing any major maize variety is the income foregone from other crops or other sources of income. In the communities of Tlaltetelco and López Mateos, tomatoes or onions are a much better business than maize, although economic risk could explain the permanence of maize. At Tecajec and López Mateos, sorghum is a better business than maize, with lower production costs and little economic risk. At Ayapango, non-farm households earn more than twice the gross income of farm households, averaging Mx$ 18,000 and Mx$ 8,000 per year, respectively. City-based income is more likely to pose the real threat to the conservation of traditional varieties in Ayapango. These findings suggest that maize-growing farm households are operating with objectives more complex than profit maximization.

7.7. CONCLUSIONS

Modern varieties have been perceived as a "threat" to crop diversity, based on the underlying assumptions that modern varieties are superior in yield and profitability and that all farmers share the sole objective of maximizing expected profits. This study raises doubts about both assumptions. First, in this area of maize diversity, data indicate that modern varieties have no clear advantages over traditional varieties in terms of yield or net income per hectare. In fact, most of the major maize types grown in the study communities are comparable in either mean yield, economic returns, or both. The opportunity cost of growing traditional maize is related to the income foregone from other crops, rather than from modern maize varieties.

If farmers considered only mean yield, this finding implies that they would be indifferent about which major maize types they choose to grow, including hybrids. In other words, they would grow any one of the major maize types, on all of their maize area. A neoclassical decisionmaking model based on maximization of expected profits cannot, therefore, explain why the farmers in these communities grow more than one maize type. This result, combined with the fact that the survey farmers grow maize even though it generates little or negative income, supports the hypothesis that they maximize utility instead of profit. Alternative economic explanations include risk and market imperfections for products, related inputs, or product characteristics. Descriptive data do not suggest that these other factors play a strong role in the choice among major maize types, but whether or not any of them is jointly or individually significant would require the development and testing of a fuller model.

7.8. IMPLICATIONS FOR LOCAL CONSERVATION OF GENETIC RESOURCES

The finding that in the study communities modern varieties have no advantage over traditional varieties in terms of mean yield or mean net income per hectare can be interpreted in one of two ways. First, as suggested by Frankel, Brown, and Burdon (1995: 84), "it may well be the case that . . . it is not the superiority of landraces which is the cause of their retention, but the lack of local plant breeders who could produce 'improved varieties' in tune with the agroecosystem." Alternatively, it may be that farmers working with broad genetic pools and natural evolutionary processes are efficient in generating competitive landraces tuned to their own agroecosystems and the needs of their society. The necessary conditions for this second hypothesis are those understood as typifying a center of crop domestication: broad genetic pools, large numbers of farmers, a large number of environments, and an essential historic food staple.

If the first hypothesis is correct, it is disturbing for a crop as important as maize in Mexico. López-Pereira and Morris (1994) calculated that roughly 23% of the maize area in Mexico is planted to modern varieties, of which about one-half are hybrids. Lack of incentives for a well-functioning seed system may also explain this outcome (Morris, 1998), but Ayapango is located near Mexico City and four major research institutes are located within 50 km.[3] Research institutes began working in the region where the study communities are located during the mid-1940s, and they have apparently released several successful modern varieties since the 1950s (Stakman, Bradfield, and Mangelsdorf, 1967; Angeles, 1968).

Neither of the hypotheses can be refuted in the context of this study, but both have important implications for *in situ* conservation. Qualset *et al.* (1997) have proposed that improvements in farmers' breeding procedures can provide an incentive to locally based conservation (see also chapter 15). For example, local varieties could be used as the basis for improvement and for introgressing genes from exotic sources. The key issue may be that *in situ* conservation can be viable *only if* farmer-based evolution of crops continues effectively.

Finally, strategies for conserving traditional varieties *in situ* should take into account the distinct nature of processes that may be involved with major and minor crop types. The most appropriate support for major types might be improvements in farmer's breeding procedures, without particular concern for the introduction of new genes or loss of alleles. The area in major varieties could easily be monitored and their possible displacement predicted. Actions could be taken if threatened varieties are important from the genetic resources point of view. For minor varieties, locally based seed-saver schemes might be more effective (Worede, 1997; Magnifico, 1996). To reduce inbreeding depression, it might be possible to arrange for the supply of intra-community composite seed of similar samples.

In summary, the agronomic and economic competitiveness of farmers' own varieties with the modern varieties that have been introduced into the study areas raises hypotheses about the efficacy of the formal seed system, including breeding programs and the seed market, relative to local farmers' systems. This finding has implications for the design of strategies to support *in situ* conservation and the costs of such strategies.

Acknowledgments

The research described in this chapter was supported through a grant of the McKnight Foundation, and the Ph.D. studies of Hugo Perales were supported by CONACYT, Mexico. Fernando Castillo G., Rafael Ortega P., Tarscicio Cervantes S., and J. M. Hernández C. classified the maize races. Alejandro Durán F., Ramiro Pérez M., Gabriel Fernández F., and Adrián Rosas S. assisted with the survey. Adrián Rosas S. also assisted with the trial. We appreciate their assistance.

Notes

1 Although the Ancho race is not characterized in Wellhausen, Roberts, and Hernández X. (1951), and Benz' (1986) description does not correspond precisely to the maize type collected in this study, the general consensus of maize experts in Mexico is that Ancho should be classified as a race and that the type collected corresponds to that race (F. Castillo G., R. Ortega P., T. Cervantes S., and J. M. Hernandez C., personal communications).

2 This so-called hybrid does not segregate as common hybrids do and may have been an open-pollinated modern variety. In any case it is clearly an introduced type that has been managed and selected as a traditional variety.

3 INIFAP, Chapingo Autonomous University, the Postgraduate College at Montecillos, and CIMMYT.

References

Angeles A., H. H. 1968. El maíz y el sorgo y sus programas de mejoramiento genético en México. In Sociedad Mexicana de Fitogenética (SOMEFI), *Memoria del Tercer Congreso Nacional de Fitogenética, I Simposio, Utilización de la Genética y la Estadística en el Mejoramiento de los Cultivos Agrícolas*. Chapingo: SOMEFI.

Bellon, M. R. 1996. The dynamics of crop infraspecific diversity: A conceptual framework at the farmer level. *Economic Botany*. 50: 26–39.

Bellon, M. R., and S. B. Brush. 1994. Keepers of maize in Chiapas, Mexico. *Economic Botany*. 48:196–209.

Bellon, M. R., and J. E. Taylor. 1993. Folk soil taxonomy and the partial adoption of new seed varieties. *Economic Development and Cultural Change* 41: 763–786.

Benz, B. F. 1986. Taxonomy and evolution of Mexican maize. Ph.D. thesis, University of Wisconsin, Madison, Wisconsin.

Brush, S. B. 1989. Rethinking crop genetic resource conservation. *Conservation Biology*. 3: 19–29.

Brush, S. B. 1995. *In situ* conservation of landraces in centers of crop diversity. *Crop Science*. 35: 346–354.

Byerlee, D. 1996. Modern varieties, productivity, and sustainability: Recent experience and emerging challenges. *World Development* 24: 697–718.

Cervantes S., T., and Mejía, A. 1984. Maíces nativos del área del Plan Puebla: Recolección de plasma germinal y evaluación del grupo tardío. *Revista Chapingo* 9 (43–44): 64–71.

CIMMYT (International Maize and Wheat Improvement Center). 1969. *The Puebla Project 1967–1969*. Mexico, D.F.: CIMMYT.

CIMMYT (International Maize and Wheat Improvement Center). 1973. *El Plan Puebla, Siete Años de Experiencia: 1967–1973*. Mexico, D.F.: CIMMYT.

Dennis, J. V. 1987. Farmer management of rice variety diversity in northern Thailand. Ph.D. thesis, Cornell University. University Microfilms International (UMI) no. 8725764. Ann Arbor: UMI.

Feder, G. 1980. Farm size, risk aversion, and the adoption of new technology under uncertainty. *Oxford Economic Papers* 32: 263–283.

Feder, G., R. Just, and D. Zilberman. 1985. Adoption of agricultural innovations in developing countries: A survey. *Economic Development and Cultural Change* 30: 59–76.

Ford-Lloyd, B., and M. Jackson. 1986. *Plant Genetic Resources: An Introduction to Their Conservation and Use*. London: Edward Arnold.

Frankel, O. H. 1970a. Genetic conservation in perspective. In O. H. Frankel and E. Bennett (eds.), *Genetic Resources in Plants: Their Exploration and Conservation*. International Biological Program (IPB) Handbook No. 11. Oxford: IPB and Blackwell Scientific Publications.

Frankel, O. H. 1970b. The genetic dangers of the Green Revolution. *World Agriculture* 19(3): 9–13.

Frankel, O. H., A. H. D. Brown, and J. J. Burdon. 1995. *The Conservation of Plant Biodiversity*. Cambridge, UK: Cambridge University Press.

Garcia, E. 1973. *Modificaciones al Sistema de Clasificación Climática de Köppen*. Mexico, D.F.: Universidad Autónoma de México (UNAM).

INEGI (Instituto Nacional de Estadística, Geografía e Informática). 1994a. *Mexico. Resultados Definitivos. VII Censo Agrícola Ganadero*. 3 vols. Mexico, D.F.: INEGI.

INEGI (Instituto Nacional de Estadística, Geografía e Informática). 1994b. *Morelos. Resultados Definitivos. VII Censo Agrícola Ganadero*. Mexico, D.F.: INEGI.

Johnson, A. W. 1972. Individuality and experimentation in traditional agriculture. *Human Ecology* 1: 149–159.

López-Pereira, M. A., and M. L. Morris. 1994. *Impacts of International Maize Breeding Research in the Developing World, 1966–1990*. Mexico, D.F.: CIMMYT.

Louette, D., A. Charrier, and J. Berthaud. 1997. *In situ* conservation of maize in Mexico: Genetic diversity and maize seed management in a traditional community. *Economic Botany*. 51: 20–38.

Magnifico, F. A. 1996. Community-based resource management: CONSERVE (Philippines) experience. In L. Sperling and M. Loevinsohn (eds.), *Using Diversity: Enhancing and Maintaining Genetic Resources On-Farm, Proceedings of a Workshop Held on 19–21 June, 1995, in New Delhi*. New Delhi: International Development Research Centre (IDRC).

Meng, E. C. H. 1997. Land allocation decisions and *in situ* conservation of crop genetic resources: The case of wheat landraces in Turkey. Ph.D. thesis, University of California, Davis, California.

Morris, M. (ed.). 1998. *Maize Seed Industries in Developing Countries*. Boulder: Lynne Rienner.

Muñoz O., A., A. Carballo C., and V.A. González. 1976. Mejoramiento del maíz en el CIAMEC. I. Análisis crítico y reenfoque del programa. In Sociedad Mexicana de Fitogenética (SOMEFI), *Memoria VI Congreso Nacional de Fitogenética*. Monterrey: SOMEFI.

Perales R., H. 1998. Conservation and evolution of maize in Amecameca and Cuautla Valleys of Mexico. Ph.D. thesis, University of California, Davis, California.

Perrin, R. K. 1975. Economic analysis of CIMMYT on-farm maize trials in Veracruz, 1973–1975. CIMMYT internal document. Mexico, D.F.: International Maize and Wheat Improvement Center (CIMMYT).

Perrin, R. K. 1976. Maize technology and its adoption in Veracruz, Mexico. CIMMYT internal document. Mexico, D.F.: International Maize and Wheat Improvement Center (CIMMYT).

Qualset, C. O., A. B. Damania, A. C. A. Zanatta, and S. B. Brush. 1997. Locally based crop plant conservation. In N. Maxted, B. V. Ford-Lloyd, and J. G. Hawkes (eds.), *Plant Genetic Conservation, the In Situ Approach*. London: Chapman and Hall.

Richards, P. 1986. *Coping with Hunger, Hazard and Experiment in African Rice-farming Systems*. London: Allen and Unwin.

Sanders, W. T., J .R. Parsons, and R. S. Stanley. 1979. *The Basin of Mexico: Ecological Processes in the Evolution of a Civilization*. New York: Academic Press.

Smale, M., and P. W. Heisey. 1998. Grain quality and crop breeding when farmers consume their grain: Evidence from Malawi. In R. Rose, C. Tanner, and M. A. Bellamy (eds.), *Issues in Agricultural Competitiveness: Markets and Policies*. International Association of Agricultural Economists (IAAE) Occasional Paper No. 7. Vermont: Ashgate Publishing Company.

Smale, M., R. E. Just, and H. D. Leathers. 1994. Land allocation in HYV adoption models: An investigation of alternative explanations. *American Journal of Agricultural Economics*. 76: 535–546.

Stakman, E. C., R. Bradfield, and P. C. Mangelsdorf. 1967. *Campaigns Against Hunger*. Cambridge, USA: Belknap Press.

Wellhausen, E. J., L. M. Roberts, and E. Hernández X. 1951. *Razas de Maíz en México*. Folleto Técnico No. 5. Mexico, D.F.: Oficina de Estudios Especiales, Secretaría de Agricultura y Ganadería.

Worede, M. 1997. Ethiopian *In Situ* Conservation. In N. Maxted, B. V. Ford-Lloyd and J.G. Hawkes (eds.), *Plant Genetic Conservation: The In Situ Approach*. London: Chapman and Hall.

Zanatta, A. C. A., M. Keser, N. Kilinç, S. B. Brush, and C.O. Qualset. 1996. Agronomic performance of wheat landraces from western Turkey: Bases for *in situ* conservation practices by farmers. In *Fifth International Wheat Conference*, June 10–14, 1996, *Ankara, Turkey. Book of Abstracts*. Ankara: Ministry of Agriculture and Rural Affairs (Turkcy), International Maize and Wheat Improvement Center (CIMMYT), International Center for Research in Dry Areas (ICARDA), Oregon State University, US Agency for International Development (USAID), and German Agency for Technical Cooperation (GTZ).

8 IMPLICATIONS FOR THE CONSERVATION OF WHEAT LANDRACES IN TURKEY FROM A HOUSEHOLD MODEL OF VARIETAL CHOICE

E. C. H. Meng, J. E. Taylor, and S. B. Brush

8.1. INTRODUCTION

The implementation of *in situ* conservation methods in conjunction with *ex situ* conservation efforts has been increasingly promoted and accepted in recent years (Maxted, Ford-Lloyd, and Hawkes, 1997). Despite the potential advantages of conserving traditional varieties on farms, however, a large gap remains between the observation of *de facto* conservation by farmers in areas of crop diversity and the establishment of a viable, long-term framework for on-farm conservation. *De facto* conservation refers to the decision by some farmers to continue cultivating landraces, even though modern varieties are available to them. The evidence of *de facto* conservation found in a number of detailed studies in areas of crop diversity (Brush, 1992; Dennis, 1987; chapter 7) is promising but provides insufficient assurance that society can rely indefinitely on farm households to conserve these resources. Nor have the existing data provided concrete suggestions regarding how such a conservation program might be implemented.

In conceptualizing an on-farm conservation program, fundamental considerations include both the characteristics of the households that are most likely to grow the varieties identified as important genetic resources and the relative importance of the factors that drive their decisions about which variety or varieties to grow. This information allows the estimation of the probabilities that households will continue to choose to cultivate these genetic resources and provides the means to assess the effects of policies on household behavior. Moreover, the probabilities provide information with which to target specific households or household groups. To understand the relationship of varietal choice decisions to on-farm conservation of genetic resources, we need also to establish a linkage with the genetic diversity that results from those choices at the household level.

This chapter discusses findings from a behavioral model analyzing the incentives influencing a household's decision to grow traditional wheat varieties and estimates factors affecting the diversity outcomes observed for those varieties. The linkage between a household's choice of variety and diversity outcomes is also tested. The data used come from a socioeconomic survey of 287 households in a three-province area of wheat diversity in Turkey. The surveyed provinces of Eskisehir, Kutahya, and Usak are located to the west of the Anatolian Plateau, one of Turkey's major wheat-producing regions. Data for calculating diversity outcomes were obtained from seed collected in farmers' fields and in household farm stores. Data from field trials with the collected seed are used to measure the diversity outcome scientifically as the morphological variation in selected wheat characteristics.

8.2. BEHAVIORAL LINKAGES IN THE TURKISH SETTING

Behavioral linkages between choice of variety and genetic diversity can be modeled as either recursive or simultaneous. Linkages are recursive, or sequential, if household land-use decisions are made independently of any diversity-outcome considerations by the farmer; that is, if the farm household does not value diversity independently for its own sake, but rather for its contribution to household production and consumption. Because field visits and farmer interviews in the surveyed area in Turkey, as in many other areas of crop genetic diversity, indicated overwhelmingly that a conscious conservation of genetic diversity for its own sake was not a household objective in selecting which wheat variety to grow, the linkage with diversity in this research is modeled as recursive.[1] It is thus primarily the household's choice of varieties that conditions diversity outcomes. Diversity outcomes are realized once the cultivation decision has been taken and the variety planted. Below, the household varietal choice decision is discussed. Next, a model for the diversity outcome is specified and estimated, and the linkage between the two tested. In the final section, we discuss the policy implications of our results.

8.3. THE HOUSEHOLD VARIETAL CHOICE DECISION

Theories of farmers' behavior under risk (Fafchamps, 1992; Finkelshtain and Chalfant, 1991; Just and Zilberman, 1983; Feder, 1980), access to markets (Fafchamps, 1992; Goetz, 1992; de Janvry, Fafchamps, and Sadoulet, 1991), and environmental constraints

(Bellon and Taylor, 1993; Jansen, Walker, and Barker, 1990; Perrin and Winkelmann, 1976) have all been used to explain the resource allocation decisions of farm households. Each of these theories alone can explain, to some extent, the selection of crop varieties. Because the omission of any one of them would likely bias the conclusions, we have incorporated them into a single, general model to provide some indication of the relative importance of the factors driving household varietal choice decisions. The additional clarification and reliability of the results provided by a general model is critical in assessing the policy alternatives that influence the conservation of crop genetic resources by farmers in areas of diversity.

The land allocation decision of the household involves a discrete choice in the household's decision to cultivate or not to cultivate a variety and a continuous choice in the amount of total household area to plant in the variety. We focus on the first, and arguably more important, of these decisions for the assessment and planning of on-farm conservation. Since the decision to cultivate and, by association, to ensure the existence of traditional varieties takes place at the household level, it is these household decisions that ultimately determine whether a landrace will be available for contributions to future crop development. The factors influencing the household's choice of variety on a given plot of land thus also influence the probability that the landrace will continue to be part of the farm system and the larger ecosystem in the future.

The household's decision is modeled as the choice between the cultivation of a traditional and modern variety for each of its available plots of land.[2] The random utility model underlying the household's choice of variety assumes that the choice reflects the household's decision in favor of the variety that provides the highest utility level of all the available alternatives. The household chooses to cultivate a variety i in plot l if the expected utility resulting from this decision is greater than the expected utility from any other choice of variety. Utility can be expressed as:

$$(8.1) \qquad U_{il} = Q'\alpha + Z'\beta + e_{il}$$

where Q is a vector of the means and variances of each variety option by plot and Z is a vector of household and market variables that affect utility directly regardless of the variety cultivated. The probability that variety i will be chosen is thus equivalent to the probability that the utility from planting variety i is greater than the utility from planting variety j.

The potentially influential variables in the household's choice of variety for a given plot of land can be divided into two groups: those with an indirect effect on choice of variety and those with a direct effect. The variables acting indirectly on choice of variety do so by shaping the household-specific yield expectations for each variety that the household considers for cultivation in the plot. Since the production outcomes of both the traditional and modern varieties are stochastic, a risk-averse household weighs the expected plot-level yield against variance of yield for each variety in making its decision. Moreover, a household concerned with risk considers its ability to withstand an unfavorable production outcome in the plot through production elsewhere on the farm. Plot-level agroecological conditions, such as soil fertility and availability of irrigation, play a central part in variety-specific outcomes, since the yield performance of each variety varies with the characteristics of the variety, the characteristics of the plot, and the interaction between the two. All else being equal, both a high level of soil fertility and the availability of irrigation in the plot are expected to have a positive effect on expected yields from modern varieties and thus a negative effect on the likelihood that the household will choose a traditional variety. In contrast, a traditional variety that is better adapted to local agroclimatic conditions may be more successful than its modern counterpart on a plot of low fertility or with no means of irrigation.

Other factors such as the availability of capital and labor inputs also indirectly affect the choice of variety through household yield distributions. Likewise, variables representing road quality and the distance to the closest market center are included to reflect the extent of difficulties in obtaining other necessary inputs such as seed, fertilizer, and information for the cultivation of modern varieties.

Effects of the explanatory variables on the variance of yield are not restricted to the same direction as yield (Just and Pope, 1979) and are largely ambiguous a priori. However, the availability of inputs such as fertile soil, irrigation, and adequate labor supply is expected to have a negative effect on variation in yield, regardless of whether the variety is a traditional or modern one. Because of the increased difficulties they create for obtaining necessary inputs, poor roads and long distances may increase the estimated variance for modern varieties, although not necessarily for traditional ones.

The second group of explanatory variables is hypothesized to have a direct effect on the probability that a household selects a particular variety for one of its plots. Several of these variables are linked to risk but do not affect the variety decision through yield expectations. Household-specific characteristics that represent the household's ability to handle risk, as reflected in the amount of land owned, the household's ability to purchase amenities and equipment, and the availability of off-farm income, are of importance in the choice of variety. A household's decision about which kind of variety to plant is also affected by its unique demand for characteristics associated with traditional varieties and its ability to obtain these characteristics in markets. For households participating in the wheat market, market prices received for different wheat varieties are a consideration in the planting decision. Market prices do not play an important role, however, for households that do not participate in the market, e.g., those facing high transactions costs in accessing markets (de Janvry and Sadoulet, 1994; de Janvry, Fafchamps, and Sadoulet, 1991).

8.4. FINDINGS FROM THE HOUSEHOLD VARIETAL CHOICE DECISION

Estimation of the household's plot-level decision proceeds in three steps: a reduced form estimation of the household's decision that includes all explanatory variables; an estimation of predicted yields and variances for each variety alternative on every household plot, correcting for selection bias and using only the variables with indirect influence on choice of variety; and a structural probit incorporating the predicted values of the yield distribution's moments as well as variables with direct influence. A detailed discussion of these estimation procedures and results can be found in Meng, Taylor, and Brush (1998).

Findings from the structural probit estimation indicate that the greater the predicted yield advantage of the traditional variety over the modern variety on a given plot, the more likely a household is to plant a traditional variety on that plot. Yield is thus an important consideration in the household's decision regarding which type of variety to plant. Nevertheless, yield advantages are tempered by the household's consideration of yield risk. The results also suggest that households take into account the linkages of plot production with expected production from elsewhere on the household farm.

Plot-level land quality characteristics appear to influence the choice of variety significantly through their effect on plot-specific yield distributions. However, having controlled for the production-related aspects of land quality, the results demonstrate a significant and positive influence from household aggregate holdings of different land qualities on the likelihood of choosing a traditional variety. Thus, environmental variables appear to shape land allocations not only through their indirect effects on plot-specific yields and variances, but also through their direct effect on the household's perception of and ability to sustain risk. Moreover, the positive effect from the ownership of fertile land supports the possibility that cultivation of traditional varieties is not always necessarily a decision limited to small, poor households farming on land of marginal quality.

Finally, the findings lend support to the importance of market considerations in shaping land-use choices. District-level price differences between traditional and modern varieties are significant in the household's decision. Households respond to higher prices received for traditional varieties by planting them with a higher probability. Furthermore, households in districts characterized by a smaller percentage of market activity, perhaps in part as a result of distance or poor roads, are significantly more likely to grow traditional varieties. Household size and livestock production, independent of production concerns, also appear to play a role in the choice of variety at the plot level. The importance of these variables may stem from a demand for consumption and feed characteristics that are uniquely associated with traditional varieties. Age and education of the household head are also influential: households characterized by older heads of household or those with fewer years of formal education are more likely to select traditional varieties over modern varieties for cultivation, all else being equal.

8.5. MARGINAL EFFECTS OF PROBABILITIES FOR THE STRUCTURAL FORM MODEL

To compare the relative importance of the factors influencing choice of variety, we examine the effects on the probability of cultivating a traditional variety from a 1% change in the level of the explanatory variable (Meng, Taylor, and Brush, 1998). The largest effect on a household's probability of choosing a traditional variety comes from a change in the percentage of output marketed in the district. This finding suggests that any policies leading to additional market infrastructure and development could have a large impact on the type of variety a household chooses to grow. A change of similar magnitude in the price difference between traditional and modern varieties also affects the probability of choosing a traditional variety, but with slightly less impact. The importance of relative prices implies a potential role for price policies, such as premiums for certain traditional varieties, in influencing the choice of variety. Changes in the plot-level constraints that are transmitted through the predicted moments of the yield distributions also indicate a relatively important influence on the probability of cultivating a traditional variety. Household-specific characteristics, such as age and education of the household head and the size of the household's labor force, appear to play a relatively large role in changing household probabilities of choosing a traditional variety as well, although factors such as these may be more difficult to manipulate with policy instruments. Finally, the size of the change in probability from variables reflecting land ownership (both land quality and total area owned) indicate the key role of the household's overall level of wealth in its cultivation decisions.

8.6. WHEAT DIVERSITY IN TURKISH FARM HOUSEHOLDS

An average of less than two varieties is cultivated per household in the surveyed region and, of the households planting traditional varieties, the cultivation of more than one landrace is not common. Crop diversity observed for wheat in these Turkish household farms is thus not predominantly expressed as inter-varietal diversity among several individual cultivars. The relatively small number of named varieties per household does not imply, however, that levels of diversity maintained on Turkish farms are insignificant. Rather, diversity is often expressed within the populations of landraces maintained by the household. Landraces by definition may be a composite of many different genotypes while improved varieties in contrast are bred specifically to consist of one unique genotype. The effects of farm management practices, such as seed selection and field preparation, are recognized as potentially important influences on variation within landrace populations. With morphological data that is scientifically measured, the amount of variation within the named landrace(s) cultivated by a household can be examined in detail. In this particular situation, the focus on diversity exclusively in landrace populations is reasonable, since the type of diversity being examined is the morphological variation within populations. Morphological variation may also occur within a modern variety among advanced generations grown for many years by farmers, but little is known about the extent of this variation and the factors that influence it.

8.7. ESTIMATION OF DIVERSITY OUTCOME

The diversity outcome for a landrace cultivated by a household is expressed as:

(8.2) $D_i = f(W, \zeta, e_i)$

where W is a vector of household characteristics, farm management practices, and agroecological influences affecting diversity outcomes, ζ denotes the inverse Mills ratio calculated from the structural form estimation of the household's plot-level decision between modern and traditional varieties, and e_i is a normally distributed random error.

8.7.1. Diversity Indices

Two diversity indices were calculated with data on morphological characteristics collected from the landrace seed obtained from survey households and grown out in the field experiments.[3] Since not all of the collected samples were included in the field experiments and thus not all households are represented, fewer plot-level observations (133 plots) are available for the diversity estimations than for the estimations on choice of variety. The diversity outcome using the Shannon index, which measures variation in qualitative traits, was calculated using nine wheat characteristics. Table 8.1 describes the qualitative traits used.

The second index of diversity was calculated with coefficients of variation from data on six quantitative traits. Selected descriptive statistics from individual components of this index are presented in Table 8.2. Both human actions and agroecological factors are expected to be instrumental in determining the amount of diversity measured on-farm. Furthering the understanding of the human actions that determine diversity, however, is of particular interest for the design of policies influencing diversity outcomes.

Table 8.1. Wheat Characteristics Used in Shannon Index of Diversity (n=133)

Variable	Description/qualitative category	Mean	Minimum	Maximum
Type	Durum or bread wheat	0.233	0	0.997
Awn color	No color, little color, heavily colored	0.507	0	0.914
Stem color	Dark, medium, light	0.241	0	0.498
Glume color	White, red, dark red	0.386	0	0.944
Awn type	Normal awn, small awn, no awn	0.006	0	0.259
Glume pubescence	Glabrous (no pubescence), slightly pubescent, pubescent	0.331	0	0.842
Awn/glume ratio	Much smaller, smaller, same size, bigger	0.232	0	0.652
Basal sterility	Existence of aborted spikes (no seed development at collar)	0.505	0.308	0.728
Kernel color	White, red, dark red	0.461	0	0.898

Source: Zanatta *et al.* (1996).

Table 8.2. Characteristics Used in CV Index (n=129)

Variable	Mean	Minimum	Maximum
Height	9.77	5.33	16.22
Basal sterility	55.31	15.92	131.60
Number of spikelets	19.50	7.68	39.50
Number of kernels	21.09	12.39	42.90
Spike length	10.42	4.89	19.41
Spike density	18.98	5.22	38.28

Source: Zanatta *et al.* (1996).

8.7.2. Household Determinants of Diversity

Households may control the variation in their fields in part through farm management practices. Relevant farm practices affecting diversity within landrace populations include seed selection practices such as the frequency of seed replacement and cleaning of fields used for future seed sources, the minimization of seed mixture during threshing, and field preparation to reduce weed problems. With constraints on labor availability and financial resources, households must weigh the opportunity costs of each task in selecting where best to direct their time and money. The available household labor supply determines the amount of time that can be dedicated to tasks that affect diversity outcomes. Because the loosening of labor constraints decreases the opportunity costs of these tasks, all else being equal, a negative relationship with intra-varietal diversity is expected *a priori* for the size of the labor force at the household's disposal.

Tractor ownership is also expected to decrease levels of observed diversity subsequent to the household's decision about which variety to grow. Household ownership of a tractor represents the presence of a certain degree of mechanization and market involvement on the farm that could both loosen constraints on labor availability and influence household attitudes towards the presence of variation in a cultivated variety. Similarly, the number of varieties managed on household land is included to indicate demands on the household's time.[4] The cultivation of multiple varieties also requires additional time. All else being equal, the cultivation of fewer varieties relieves labor constraints in the household. This variable is thus expected to have a negative effect on observed intra-varietal diversity.

Household wealth may decrease variation within landrace populations by enabling the use of outside labor or the purchase of other time-saving inputs. Wealthy households may also be more likely to participate in market activity and thus predisposed in their attitudes against high levels of variation in cultivated varieties, particularly if the marketed variety incurs price penalties for a mixture of wheat types. Risk considerations and the luxury aspect of traditional varieties that are linked to household wealth levels do not enter into the diversity function since they have already been accounted for in the separately estimated varietal choice decision. The household's wealth status, represented by off-farm income and the household wealth index (HWI), is expected to have a negative influence on household diversity outcomes.

Age and education level of the household head are included to reflect the importance of traditional farming methods and attitudes in the household. A farmer with more years of schooling or a farmer with a more modern outlook may be less inclined to accept the presence of a large amount of variation in his/her fields, again perhaps because of market influences. Conversely, an older farmer may be more likely to be unconcerned about market considerations. Negative relationships between these variables and observed diversity are expected.

The use of price as a potential policy tool for ensuring targeted levels of genetic diversity is addressed with the inclusion of two market-related variables, average district price for traditional varieties and the percentage of participation in markets at the district level. Government-determined market prices for wheat are currently based on wheat type (e.g., bread or durum) and classification (e.g., Hard White, Hard Red, and others).[5] No official distinction is made between modern and traditional varieties. Standards and grades are well established, and farmers are penalized monetarily for the presence of foreign material and excessive mixture of types. By the time the wheat arrives at the market, much of the variation observable in the field (e.g., plant height, presence or absence of awns) may no longer be visible; however, variation in traits such as seed color and wheat type remain easy to detect. Average district prices for traditional varieties are used to test the hypothesis that lower levels of diversity command higher market prices.

The percentage of wheat output marketed by district is included to reflect the fact that many of the surveyed households do not participate in market activities. Districts characterized by low percentages of marketed output will not be affected by price incentives as much as districts with a more active role in wheat markets. The variable also provides an indication of the extent to which diversity levels on the farm are influenced by market activity.

8.7.3. Agroecological Determinants of Diversity

Agroecological heterogeneity is instrumental in fostering the existence of variation in crop characteristics due to the creation of different production niches and unique sets of selection pressures (Bellon, 1996). Stresses caused by inadequate rainfall and soil moisture, excessively high and low temperatures, and potential evaporation all have the potential to shape wheat development and variation (Loss and Siddique, 1994; Pecetti, Damania, and Kashour, 1992). Variables classifying households by province are thus included to reflect differences in cultivation environment among the provinces.[6] Since the influence of geographical location on household varietal choice decisions has already been taken into account, this effect should not be confused with the separate effect of provincial location on diversity outcomes. Provincial variables may also pick up differences among intangible characteristics unique to each province.

The elevation of each village is included to reflect differences in growing conditions not picked up by provincial classifications. Elevation is a potentially important factor in explaining diversity outcomes since it incorporates information on a number of influential natural factors, such as air and soil temperature, precipitation levels, the severity of the winters, and air pressure. In previous studies, elevation has been found to have a positive effect on observed diversity up to very high elevations (Engels, 1994). Table 8.3 shows the descriptive statistics for the variables used to estimate diversity indices.

Heterogeneity in the growing environment, hypothesized to be positively associated with diversity levels, is represented in the diversity function by the extent to which the household's land is subdivided into plots. The fragmentation index is calculated as the ratio of household plots to the total area of land owned.[7] High levels of fragmentation may also reflect the possibility that household land is not contiguous. Distance between a household's plots could potentially affect the heterogeneity of the household's growing environment by increasing the number of production niches (Brush, Taylor, and Bellon, 1992). Although the area of individual plots may decrease as land fragmentation increases, thereby increasing the uniformity within plots, for a household of a given size, the overall heterogeneity in the cultivation environment is likely to be higher when the farm has many separate plots. The total number of farmer-reported soil types present on the household farm is also included in the estimation.[8]

Table 8.3. Variables in Estimation of Diversity Function (n=133 plots)

Variable	Mean	Minimum	Maximum
Agroecological determinants of diversity			
Kutahya	0.65	0	1
Usak	0.32	0	1
Village elevation (m)	1,085.83	700	1,450
Number of soil types	1.44	1	3
Fragmentation index	0.07	0.004	0.25
Human determinants of diversity			
Total number varieties cultivated by household	1.23	1	3
Age of household head	47.30	18	85
Education of household head (no. of yr)	4.38	0	13
Household labor (no. people aged 13 or older)	5.05	2	8
Tractor ownership (1=yes, 0=no)	0.67	0	1
Off-farm income (1=yes, 0=no)	0.22	0	1
Household wealth index	0.43	0	1
Average district traditional variety price	724.81	594.50	769.78
Percentage of district production sold	0.43	0.18	0.65

8.8. TESTING THE LINKAGE BETWEEN CHOICE OF VARIETY AND THE DIVERSITY OUTCOME

As defined in this study, the household will not maintain any landrace-based diversity if it does not cultivate landraces. However, once a decision in favor of a traditional variety has been taken, several potential factors influence the level of the diversity maintained within the population. Factors not captured in the explanatory variables may relate the household's choice of variety to the diversity outcome. The nature of the relationship between the household's varietal choice decision and its diversity outcome holds potentially important policy implications for *in situ* conservation. Including in the diversity outcome estimation the inverse Mills ratio, ζ, from the structural probit estimation of the household's plot-level variety decision provides a means of testing the independence of this relationship. Rejection of the null hypothesis that the coefficient of ζ equals zero supports the notion that the probability of growing landraces is related to intra-landrace diversity measured at the level of the household. Statistical insignificance of the coefficient would indicate that the null hypothesis of independence cannot be rejected, implying that diversity outcomes are not likely to depend on *ex ante* probabilities of growing traditional varieties. Either finding would influence decisions regarding the type of household most suitable to include in an on-farm genetic resource conservation program and the nature of policy interventions for a stated conservation goal. The presence of the variable, ζ, also statistically corrects for the non-zero expected error resulting from the lack of data regarding diversity outcomes for modern varieties.

Results from the estimation of the diversity functions using both the Shannon and CV indices are provided in Table 8.4. Two findings stand out in particular. First, the insignificance of ζ in both diversity estimations suggests that the intra-landrace diversity outcome is independent of a household's probability of choosing a traditional variety over a modern variety in the plot. Thus, households with a high *ex ante* probability of growing a traditional variety are not inherently more able or more likely to maintain high diversity outcomes than households with a low *ex ante* probability of cultivating a traditional variety. An examination of the distribution of the diversity indices for the two groups of households also shows that there is little difference between the levels of diversity maintained by households with a high probability of growing landraces and those with a low probability, for a given landrace. Therefore, in identifying households for participation in an *in situ* program, targeting households with the highest probability of cultivating a given landrace may be the only means necessary to influence diversity.

Second, it is of methodological importance that signs and levels of significance of explanatory variables differ greatly depending on the diversity index used. These findings demonstrate that the choice of diversity index can result in policy implications that are potentially quite different.

8.9. POLICY IMPLICATIONS OF FINDINGS FOR ON-FARM GENETIC RESOURCE CONSERVATION

The results of the household varietal choice model, the model of diversity outcomes, and the test of relationship between the two contribute to policy efforts for crop genetic conservation by providing information on the types of households to target. They also help to identify potential geographical areas for *in situ* programs, should a more active

Table 8.4. Estimation Results of Shannon and CV Index Diversity Functions

Variable	Shannon			CV		
	Estimated coefficient	Standard error	t-statistic	Estimated coefficient	Standard error	t-statistic
Constant	0.059	0.101	0.51	22.294	8.715	2.56
Inverse Mills ratio	−0.005	0.027	−0.20	−3.285	2.285	−1.44
Kutahya	0.092	0.042	2.17	−7.724	3.613	−2.14
Usak	0.191	0.054	3.53	−6.335	4.584	−1.38
Elevation	0.00008	0.00005	1.65	0.001	0.004	0.29
Age	0.001	0.0008	1.35	0.031	0.071	0.43
Years of education	0.006	0.004	1.40	−0.001	0.382	−0.001
Household labor	−0.017	0.004	−3.78	−0.183	0.391	−0.47
Tractor ownership	−0.01	0.019	−0.51	3.309	1.635	2.02
Household wealth index	−0.069	0.018	−3.91	0.019	1.538	0.01
Off-farm income	−0.039	0.019	−2.08	0.675	1.659	0.41
Fragmentation index	−0.0241	0.134	−1.80	10.130	11.562	0.88
Soil classes	−0.013	0.013	−1.00	1.550	1.150	1.35
Household varieties	0.004	0.013	0.33	2.440	1.096	2.23
District output marketed × district traditional price	0.0005	0.0001	3.45	−0.008	0.012	−0.68
R^2	0.34			0.16		
Adj. R^2	0.26			0.05		

role become necessary in encouraging on-farm conservation. As suggested by the results of the test on the linkage between choice of variety and diversity outcomes, the policy interventions for intra-landrace diversity may be distinct from those for inter-varietal diversity. Below, we explore the implications of the research for inter-varietal choice, and hence, the probability of maintaining landraces as a source of diversity.

8.9.1. Targeting Households

The model of household choice provides the means of estimating the probabilities that households will continue to grow landraces and the effects on these probabilities resulting from changes in the factors that influence the cultivation decision. These probabilities figure largely in the identification of households most likely to grow landraces and in the minimization of the costs of encouraging farmer participation. They are the key elements in determining where the attention for on-farm conservation is most appropriately focused, in terms of both specific households and regions.

Ensuring a targeted level of diversity does not necessarily imply a preservation of the *status quo* for the households currently cultivating traditional varieties. Since much of the range in variation observed in a landrace is similar from one household to another, it is not necessary that all households currently cultivating traditional varieties continue to do so. The estimated probabilities from the model of varietal choice can be used to identify households with the highest *ex ante* probability of cultivating traditional varieties. Because these are the households that are most likely to continue growing traditional varieties of their own accord, they are also most likely to be the households for which incentives for diversity can be created and/or maintained at the lowest cost. An examination of these households reveals that they are not necessarily the poorest households, nor are they the households with the most marginal growing conditions. Because of the many positive characteristics associated with traditional varieties, particularly with respect to consumption, it is quite possible that positive wealth effects exist for diversity levels.

Table 8.5 presents selected characteristics of surveyed households characterized by their predicted probability of cultivating traditional varieties. Column 1 presents average characteristics of the surveyed households for which the estimated probability of cultivating traditional varieties was over 95%, while column 2 presents the same information for households whose probabilities of choosing traditional varieties were estimated at below 5%. The mean values of these characteristics over all households in the sample are given in column 3.

The coefficient of variation of yield for traditional varieties in the high- and low-probability households is almost identical with values of 0.25 and 0.24, respectively; however, the coefficient of variation for modern varieties for these households at 0.18 and 0.13, respectively, presents a marked contrast. Based on the predicted moments of the yield distribution, the choice between traditional and modern yields for households with the lowest probability of choosing traditional varieties clearly favors the choice of a modern variety. Moreover, the necessity for one variety to satisfy multiple requirements is probably not as crucial for this subset of households as for the other surveyed households. With a coefficient of variation of 0.25 predicted for traditional varieties and a coefficient of variation of 0.18 predicted for modern varieties, predicted yield distributions for the households with high probabilities do not completely explain the choice of traditional varieties. Given that households in general are faced with multiple requirements to satisfy, it is once again apparent that yield and yield variance, while important, are not necessarily decisive factors for many households. That yield and yield risk are not the only factors considered in the varietal choice decision is further reinforced by a comparison of the high-probability households with the average household surveyed. Coefficients of variation for both traditional and modern varieties are nearly identical, yet the probability of choosing a traditional variety is quite different. Other factors clearly influence the probability of a household's cultivation decision in favor of traditional varieties.

8.9.2. Number and Geographical Distribution of High-Probability Households
The geographical distribution of the most promising households to target for participation in an *in situ* program also needs to be considered. The location of households cultivating traditional varieties in a limited geographical area poses the potential risk that these households may become exposed to a common crop disease, pest, or other event affecting all households. A wide geographical area that incorporates a large number of agroecological niches is more likely to provide the opportunity for a greater range of variation, and thus the potential for more diversity, than a narrowly defined geographical area. Table 8.6 provides the geographical distribution of household plots with probabilities of cultivating traditional varieties estimated at above 95%.

Households located in both of the districts surveyed in Usak Province constitute an overwhelming majority of the households with the highest probability of cultivating traditional varieties. All of the 46 households surveyed in Banaz cultivated traditional varieties, while 43 of the 46 households surveyed in the Merkez district also chose at

Table 8.5. Selected Characteristics by Probability of Cultivating Traditional Varieties

Variable	High-probability households (probability >95%)	Low-probability households (probability <5%)	All households
No. households (no. plots)	111 (279)	44 (220)	284 (1,264)
Predicted yield, traditional variety (kg/dec)	189.4	170.8	190.7
Predicted variance, traditional variety	2,210.7	1,705.5	2,042.3
CV, traditional variety	0.25	0.24	0.25
Predicted yield, modern variety	235.8	321.8	271.7
Predicted variance, modern variety	1,840.5	1,646.2	2,353.3
CV, modern variety	0.18	0.13	0.18
Total land owned (ha)	10.0	27.0	16.0
Livestock (no. head)	3.5	4.2	4.0
No. in household 13 years or older	4.7	3.8	4.5
Distance to market (km)	17.3	15.3	16.4
District-level avg. traditional var. price (TL/kg)	700.3	631.3	678.8
District-level avg. modern var. price (TL/kg)	606.4	735.0	699.0
Percentage district output sold	28.0	74.0	0.56
Household wealth index (HWI=1)	0.38	0.74	0.58

least one traditional variety. In terms of total number of plots, traditional varieties were planted in 94–95% of all plots in both districts. Usak households are followed by households, particularly those located in hillside and mountain regions, in the Cavdarhisar district of Kutahya. Eighteen of forty-seven households, or 38%, cultivated at least one traditional variety on 18% of the total plots planted in the district. While there are a number of plots with high estimated probability of landrace cultivation in mountain households in the Seyitgazi district of Eskisehir Province, the number of surveyed households involved is small. Only 4 out of 49 households surveyed (6% of available plots) were involved in the cultivation of traditional varieties.

An examination of the distribution across agroclimatic zone shows that the highest probability group of households make up 12% of the total number of surveyed valley households, 33% of hillside households, and 51% of mountain households.

Since the range in variation observed in households among subpopulations of wheat landraces is less than the amount of variation among landrace populations, the marginal contribution of an additional household to subpopulation diversity may potentially be small. However, to ensure the continuing evolution of landraces, which is one of the primary advantages associated with *in situ* conservation, a network of households should be maintained in such a way that seed exchange and other community-level activities continue to take place. The involvement of additional households in the cultivation of the landrace also provides insurance against the possibility that the landrace disappears completely.

8.10. CONCLUSION

Results from the estimation of the household behavioral model confirm the role of multiple factors in determining the household's plot-level choice of variety. The independence between the decision to cultivate a traditional variety and the level of morphological diversity maintained in that variety suggests that different policy interventions might be required depending on whether the goal is to conserve intra-landrace diversity or to encourage the continued cultivation of traditional varieties. However, if the goal is to maintain morphological diversity in the wheat populations cultivated by farmers, and a significantly greater amount of morphological diversity continues to be found among traditional than among modern varieties, then the additional information provided by the diversity estimations is helpful. The similarity in the range of variation maintained in a given landrace by both high- and low-probability households

Table 8.6. Number of Households (Plots) with >95% Probability of Growing Traditional Varieties, by Location, Turkey

District/Province	Valley	Hill	Mountain
Seyitgazi/Eskisehir	0	1 (1)	3 (18)
Cavdarhisar/Kutahya	2 (2)	11 (32)	5 (32)
Merkez/Usak	10 (22)	10 (20)	23 (48)
Banaz/Usak	12 (22)	22 (50)	12 (32)

implies that it may be sufficient to identify those households with the highest *ex ante* probabilities of cultivating traditional varieties. That group of households consists of those requiring the minimum number of external incentives for *de facto* conservation. This study presents concrete steps for monitoring, predicting, and developing potential mechanisms to encourage farmers' *in situ* conservation of crop genetic resources.

Notes

1 Unfortunately, the assumption of recursivity cannot be tested with the available data.
2 In this chapter, "modern varieties" are defined as varieties that have been scientifically improved by plant breeding programs, while "traditional" or "local" varieties are defined as those that have evolved through a combination of natural and human selection. Typically, "modern" is used to refer more specifically to semidwarf wheat varieties released by plant breeding programs, but in Turkey modern varieties include both improved tall varieties and semidwarf varieties. This terminology may also differ depending on the crop (see other chapters in this volume).
3 See Meng (1997) and Zanatta *et al.* (1996) for additional details regarding the field experiments.
4 The diversity in the dependent variable, measured by the morphological variation within a landrace (intra-varietal diversity), is a different measure of diversity than the number of varieties cultivated (inter-varietal diversity). The use of the number of household varieties as an explanatory variable does not create simultaneity problems.
5 Official classes of bread wheat, in decreasing order of price paid, are: Hard White, Anatolian Hard Red, Hard Red, Semi-Hard White, Semi-Hard Red, Soft White, and Other. There are two categories of durum wheat: Anatolian Durum and Durum.
6 The use of provincial indicator variables is not perfect since weather patterns are by no means uniform within a province. However, data from provincial meteorological stations corresponding more precisely to surveyed villages and districts were too incomplete for use.
7 Total household land is assumed to be fixed.
8 Land fragmentation may have originally taken place as a result of naturally occurring changes in soil type; however, the pair-wise correlation between the number of plots and the number of soil types at present is low. Households report considerably more plots in the household than soil types.

References

Bellon, M. R. 1996. The dynamics of crop infraspecific diversity: A conceptual framework at the farmer level. *Economic Botany* 50: 26–39.

Bellon, M. R., and J. E. Taylor. 1993. Folk soil taxonomy and the partial adoption of new seed varieties. *Economic Development and Cultural Change* 41: 763–786.

Brush, S. 1992. Reconsidering the green revolution: Diversity and stability in cradle areas of crop domestication. *Human Ecology* 20: 145–167.

de Janvry, A., M. Fafchamps, and E. Sadoulet. 1991. Peasant household behaviour with missing markets: Some paradoxes explained. *Economic Journal* 10: 1400–1417.

de Janvry, A. and E. Sadoulet. 1994. Structural adjustment under transaction costs. In F. Heidhues and B. Kerr (eds.), *Food and Agricultural Policies under Structural Adjustments*. Frankfurt, Germany, and New York: Peter Lang.

Dennis, J. V. 1987. Farmer management of rice variety diversity in northern Thailand. Ph.D. thesis, Cornell University, Ithaca, New York.

Engels, M. M. 1994. Genetic diversity in Ethiopian barley in relation to altitude. *Genetic Resources and Crop Evolution* 41: 67–73.

Fafchamps, M. 1992. Cash crop production, food price volatility, and rural market integration in the third world. *American Journal of Agricultural Economics* 74: 90–99.

Feder, G. 1980. Farm size, risk aversion and the adoption of new technology under uncertainty. *Oxford Economic Papers* 32: 263–283.

Finkelshtain, I., and J. Chalfant. 1991. Marketed surplus under risk: Do peasants agree with Sandmo? *American Journal of Agricultural Economics* 73: 557–567.

Goetz, S. 1992. A selectivity model of household food marketing behavior in sub-Saharan Africa. *American Journal of Agricultural Economics* 74: 444–452.

Jansen, H. P., T. Walker, and R. Barker. 1990. Adoption ceilings and modern coarse cereal cultivars in India. *American Journal of Agricultural Economics* 72: 653–663.

Just, R. E., and R. D. Pope, R. 1979. Production function estimation and related risk considerations. *American Journal of Agricultural Economics* 61: 276–284.

Just, R. E., and Zilberman, D. 1983. Stochastic structure, farm size, and technology adoption in developing agriculture. *Oxford Economic Papers* 35: 307–328.

Loss, S., and Siddique, K. H. M. (1994) Morphological and physiological traits associated with wheat yield increases in Mediterranean environments. *Advances in Agronomy* 52: 229–276.

Maxted, M., B. V. Ford-Lloyd, and J. G. Hawkes (eds). 1997. *Plant Genetic Conservation: The In Situ Approach.* London: Chapman and Hall.

Meng, E. C. H. 1997. Land allocation decisions and *in situ* conservation of crop genetic resources: The case of wheat landraces in Turkey. Ph.D. thesis, University of California, Davis, California.

Meng, E. C. H., J. E. Taylor, and S. B. Brush. 1998. *Technology Adoption and the On-Farm Conservation of Crop Genetic Resources.* CIMMYT Economics Working Paper. Mexico, D.F.: International Maize and Wheat Improvement Center (CIMMYT).

Pecetti, L., A. B. Damania, and G. Kashour. (1992) Geographic variation for spike and grain characteristics in durum wheat germplasm adapted to dryland conditions. *Genetic Resources and Crop Evaluation* 39: 97–105.

Perrin, R., and D. Winkelmann. 1976. Impediment to technical progress on small versus large farms. *American Journal of Agricultural Economics* 58: 888–894.

Zanatta, A. C. A., M. Keser, N. Kilinç, S. B. Brush, and C.O. Qualset. 1996. Agronomic performance of wheat landraces from western Turkey: bases for *in situ* conservation practices by farmers. In *Fifth International Wheat Conference, June 10–14, 1996, Ankara, Turkey. Book of Abstracts.* Ankara: Ministry of Agriculture and Rural Affairs (Turkey), International Maize and Wheat Improvement Center (CIMMYT), International Center for Research in Dry Areas (ICARDA), Oregon State University, US Agency for International Development (USAID), and German Agency for Technical Cooperation (GTZ).

IV

Impacts of Crop
Diversity on Productivity
and Sustainability

9 THE CONTRIBUTION OF GENETIC RESOURCES AND DIVERSITY TO WHEAT PRODUCTIVITY IN THE PUNJAB OF PAKISTAN

J. Hartell, M. Smale, P. W. Heisey, and B. Senauer

9.1. INTRODUCTION

The ability to meet the world's growing food demand dramatically improved with the release of modern semidwarf or "green revolution" wheat varieties in the early 1960s. Despite initial and subsequent success in improving yield potential and yield stability, maintaining disease resistance, and improving other plant characteristics, the green revolution has provoked criticism and debate. A major issue in this debate concerns the effect of scientific plant breeding technology on crop genetic diversity, which is believed to have important implications for global and national food security and producer welfare.

The "green revolution" in wheat refers specifically to the development and diffusion of semidwarf wheat varieties in the developing world, which began in Mexico and South Asia during the 1960s. These semidwarf varieties contain the *Rht1* and *Rht2* genes, individually or in combination. *Rht1* and *Rht2* (two of several dwarfing genes that have been found in wheat) confer a positive interaction between a wheat genotype and its environment, by which yield increases prove greater given a favorable combination of soil moisture, soil fertility, and weed control. The genes were initially introduced in Japanese breeders' materials through Daruma, believed to be a Korean landrace (Dalrymple, 1986). A cross descended from Daruma, Norin 10, was introduced into a United States breeding program at Washington State University in 1949, and the dwarf characteristics from Norin 10 were successfully incorporated into the first "green revolution" wheats by N.E. Borlaug in Mexico. The semidwarf wheats currently developed by CIMMYT and many national breeding programs in developing countries are descendants of the first green revolution varieties, but their pedigrees also contain many distinct ancestors and landraces from other sources.

Diversity, broadly considered in the biological sense, refers to the collective dissimilarity of species. Within a single crop species, diversity refers to the genetic variation that results in differing expressions of traits among individuals, which is in turn the basis of plant breeding and selection programs. Lack of diversity within a single crop species can potentially limit the ability of scientists who work with those species to respond to unknown or evolving pests, pathogens, or environmental conditions. In the case of wheat rusts, for example, widespread cultivation of varieties with a similar genetic basis of resistance increases the risk of pathogen mutation and the spread of disease once a mutation occurs. Some researchers have suggested that agricultural production systems may be similarly vulnerable, based on the assumption that modern breeding programs utilize a narrow range of genetic material, that different varieties are in fact closely related, and that genetic uniformity in breeding and production is increasing (see Frankel, 1970; National Research Council, 1972; chapters in Cooper, Vellvé, and Hobbelink, 1992).

Our perspective in this chapter derives from an interest in the effect that genetic resources and diversity have on production outcomes for farmers and, in turn, how their production choices and constraints affect diversity. The chapter begins by examining the patterns of varietal diversity in wheat as they occur in farmers' fields, using several different indicators of genetic resource use and diversity (see chapter 2 for a detailed discussion of these indicators and their meaning). Next, we use econometric estimation to analyze how varietal diversity and genetic resource type, along with other conventional inputs, contribute to wheat output and yield stability. The districts of the Punjab of Pakistan over 1979–1985 were chosen for this study because the Punjab was one of the first areas in the developing world where farmers adopted semidwarf wheat varieties. Pakistan is also one of the four largest producers of wheat among developing countries, and its per capita wheat consumption ranks among the highest.

9.2. DATA AND VARIABLES

Cross-sectional, time-series wheat production data were obtained for the six years 1979/80 to 1985/86 for each of 29 districts of the Punjab of Pakistan. Input and production data were collected from various issues of *Punjab Agricultural Statistics*, Government of Pakistan, Punjab, Lahore. As the objective was to measure the contribution of specific inputs to wheat yield over geographic areas, the unit of observation was input and output per hectare, by district.

We have expressed genetic resource use and diversity in terms of five variables derived from both social and biological science. The number of different landraces per pedigree represents the anonymous contribution of farmers' selections to the germplasm, while the number of different parental combinations expresses the contribution of plant breeding programs. These two variables represent the types of genetic resources used in the development of the varieties grown by farmers. The diversity of wheat varieties in a geographical area is indicated by a Herfindahl index of area concentration (spatial diversity; see Pardey *et al.*, 1996) and by the age of varieties grown (temporal diversity; see Brennan and Byerlee, 1991). The relative genealogical dissimilarity of varieties grown in a geographical area is measured using a distance indicator constructed from genealogical information.

The indicator of genealogical dissimilarity is based on that developed and applied by Weitzman (1992; 1993) and is given by the total branch lengths of a dendrogram or taxonomic tree constructed from cluster analysis of pair-wise distances. Each pair-wise distance is a coefficient of diversity (COD), calculated as one minus the coefficient of parentage (COP) between any two varieties. The COP estimates the pair-wise similarity of parentage among a set of varieties based on detailed pedigree analysis and has been described as measuring the transmission of latent diversity. The unweighted average and area-weighted average values of the COP or COD can be used to describe the diversity among a set of distinct varieties grown in a particular geographic area (chapter 2; Souza *et al.*, 1994; Cox, Murphy, and Rodgers, 1986).

Together, these indicators are used in the regression analysis to describe the system of wheat genetic resource use and the pattern of varietal diversity found in farmers' fields. They are defined in Table 9.1, along with dependent variables and conventional inputs considered to be important determinants of wheat output according to economic theory. In the next section, we use them to summarize patterns of genetic resource use and diversity among the varieties grown by farmers in the Punjab of Pakistan.

Table 9.1. Definition of Variables

Yield	Wheat yield in metric tons per hectare for each district and year.
Yield-DY	Wheat yield variability expressed in metric tons per hectare, using mean for each district and year, with trend removed.
Irrigation	Proportion of wheat cropped area under irrigation for each district.
Fertilizer	Total nitrogen, phosphorus, and potassium fertilizer used in wheat production, in kilograms of nutrients per hectare.
Rain	Total cumulative annual rainfall for each district in millimeters.
Literacy	Proportion of the district population that is literate.
Tractors	Availability of mechanized traction in each district, expressed as the number of tractors per hectare.
Bullocks	Availability of non-mechanized traction in each district, expressed as the number of bullocks per hectare.
Labor	Labor used in wheat production for each district. Expressed as total man-hours per hectare, where one man-day is the amount of labor performed by a healthy male working seven hours.
Landraces	Average number of different landraces per pedigree of varieties grown.
Parental combinations	Average number of different parental combinations per pedigree of varieties grown.
Genealogical distance	Measured as the total branch length of a dendrogram constructed from cluster analysis of pair-wise coefficients of diversity between varieties grown. Coefficient of diversity = 1– coefficient of parentage.
Concentration	Spatial diversity or concentration of wheat area in varieties grown. Measured as the sum of squared area shares.
Age	Temporal diversity. Measured as average age of varieties grown.

9.3. PATTERNS OF GENETIC RESOURCE USE AND DIVERSITY

9.3.1. Genetic Resource Use

During the study period 1975–1985, 18 varieties of semidwarf bread wheat were cultivated by farmers in the Punjab of Pakistan. This number excludes *desi* varieties and the tall-statured varieties that are older releases of the national wheat breeding programs. "*Desi*" varieties are traditional varieties that are selected by farmers. Varieties released from the early 1900s through the late 1960s by the Indian plant breeding program were taller than the semidwarf wheats. In fact, only one of the *desi* varieties still grown during the study period is known to have been a landrace (Roti); all others were descendants of older releases referred to as the "C–series."

Over time, the genetic resources embodied in the pedigrees of semidwarf wheat grown in the Punjab have increased both in terms of the average number of different landraces per pedigree and the average number of different parental combinations per pedigree (Table 9.2). For all districts of Punjab taken together, from the perspective of overall breeding effort, the number of different landraces is positively correlated with the number of different parental combinations in the pedigree, demonstrating that wheat breeders are using materials with new ancestors in their pedigrees. At the district level, the average number of landraces and parental combinations in the pedigrees of wheat varieties has also increased, although the magnitude differs by district, reflecting the number of varieties grown and the adoption of more recently improved varieties with longer pedigrees (Hartell, 1996).

Table 9.2. Semidwarf Wheat Varieties Grown in the Punjab of Pakistan, 1975–1985

Variety	Year of release in Pakistan	Number of different landraces in pedigree	Number of different parental combinations in pedigree	Peak area during study period (%)
Mexipak	1965	37	58	7.4
Chenab-70	1970	36	62	7.5
Blue Silver	1971	39	90	3.8
SA-42	1971	38	88	0.5
Sandal	1973	42	94	5.0
Lyallpur-73	1973	44	111	11.3
Pari-73	1973	42	94	1.1
Yecora	1974	42	94	55.6
Nuri	1975	42	94	1.4
SA-75	1975	41	71	2.6
Lu-26	1976	44	112	1.6
HD-2009	1978	37	71	0.2
Sonalika	1978	39	90	10.1
WL-711	1978	45	109	18.4
BWP79	1979	38	88	2.0
Pavon	1980	47	127	4.7
Punjab-81	1981	41	89	14.0
Pak-81	1981	49	131	9.4

Source: CIMMYT Wheat Pedigree Management System and Bureau of Statistics, Lahore, Pakistan (1986).

9.3.2. Spatial Diversity

Figure 9.1 shows the percentage distribution for all Punjab of each cultivated wheat variety and groups of varieties, or their relative abundance. Quite clearly the concentration of the most popular variety has declined in Punjab over the study period. That the dominant variety in the final period accounted for only 18% of the wheat area indicates the presence of an increasing number of varieties, each occupying a relatively small area, and is a clear indication that the spatial diversity of wheat varieties improved during the study period. Since the adoption of varieties follows cyclical patterns, however, this finding depends on the specific time period and its length.

Disaggregated district-level data show a more volatile pattern in varietal concentration over the same period (Hartell, 1996). Data show a greater concentration of a single semidwarf variety in the drier rainfed regions as well as a "boom-bust" cycle of changing concentration in more favorable production environments. The latter pattern is likely to be the result of intense disease pressure. In favorable environments, farmers must rapidly replace susceptible varieties with the high-yielding, resistant varieties.

9.3.3. Temporal Diversity

To some extent, "diversity in time" substitutes for spatial diversity in modern agriculture (Duvick, 1984). The average age of varieties grown by farmers is an expression of the rate of varietal replacement. A high average age among the varieties grown by farmers indicates that they retain the same varieties for many years. While the average age partially reflects the turnover of varieties released by the research system, the weighted average age adjusts for the effects of the distribution of varieties over a geographic area.

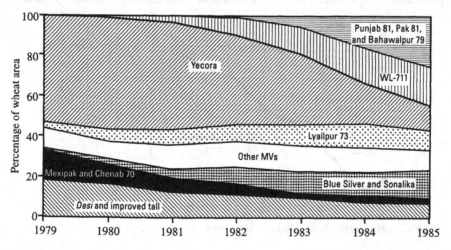

Figure 9.1. Percentage Distribution of Wheat Varieties by Area, Punjab of Pakistan, 1979–1985.

Note: Other MVs (modern varieties) include HD-2009, Lu-26, Nuri, Pari, Pavon, SA-42, SA-75, and Sandal.

Temporal diversity in the Punjab of Pakistan has been lower (i.e., the average age of varieties has been higher) than temporal diversity in many other world wheat-growing regions (Brennan and Byerlee, 1991). Province-wide data for a longer period of time, which encompasses the period covered in this study, indicate that both average age and weighted average age have increased (Figure 9.2). This reflects the persistence of older varieties and suggests that farmers have responded slowly to research output and disease pressure. The weighted average age has risen more slowly than the simple average, however, and the weighted average has been lower than the average from 1983 onward. This indicates that area planted has shifted in favor of newer wheat releases over time. Furthermore, district-level figures show that at a more disaggregated level, turnover can be considerably faster (Hartell, 1996).

9.3.4. Genealogical Dissimilarity

Changes in average and weighted coefficients of diversity, calculated from matrices of pair-wise coefficients of parentage, are presented in Figure 9.3. Unweighted average coefficients remained fairly stable at a relatively high level over the study period. The area-weighted average is very low at the beginning of the period—lower than 0.5, which is the coefficient of diversity for the parent-offspring relationship (see also Souza *et al.*, 1994). In later years, it rose to nearly 0.70.

The difference between the unweighted and area-weighted measures expresses the impact of seed distribution systems and other socioeconomic factors related to the adoption of varieties. Farmers will choose to grow the variety that is most attractive to them (in terms of profits or other measures of economic value), but their choices are limited to the seed types that are available to them. It may be important to remember that while wheat science has direct influence over the genealogies of varieties, it has little to no influence over the complex of socioeconomic factors that affect farmers' adoption decisions, the area distribution of varieties, and their replacement by farmers.

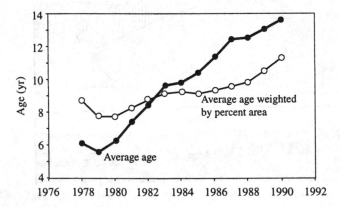

Figure 9.2. Temporal Diversity, Punjab of Pakistan, all Semidwarf Cultivars, 1978–1990.

The difference between the average and weighted coefficients of diversity varies considerably among districts, reflecting differences in farmers' objectives and in the availability and suitability of varieties to production environments. Similarly, estimates of genealogical dissimilarity based on Weitzman's index are heavily influenced by the numbers and types of varieties grown in a particular region and show large differences among districts (Hartell, 1996).

9.3.5. Summary

The evidence gathered here demonstrates a movement towards greater diversity in the group of semidwarf wheat varieties grown by farmers and their area distribution for all Punjab during the brief period under study, as measured by spatial and genealogical indicators. When measured by average age, temporal diversity appears to have declined in part because of the long-term persistence of varieties grown by farmers. By contrast, area-weighted average age indicates a more favorable rate of varietal turnover, although it is still below recommended rates of replacement. The persistence of varieties has the effect of increasing the average coefficient of diversity as well as genealogical distance to the extent that it implies a greater number of varieties with diverging pedigrees are grown.

Distinct patterns of diversity, which are likely to be highly related to the production environment and the availability of suitable cultivars, are evident from the examination of district-level data (Hartell, 1996). Variability and interactions that occur among diversity indicators at the district level are not so apparent when the analysis is conducted at the level of the province. There are two implications of these findings. First, observations of diversity at the provincial level may be a poor guide to sound policy recommendations implemented at the local level. Second, efforts to affect diversity at the farm level may require sets of policy instruments specially tailored to each production and agroclimatic environment.

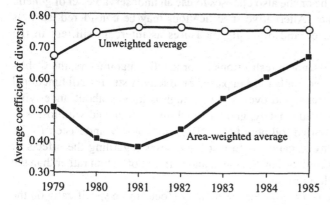

Figure 9.3. Genealogical Diversity of Wheat Varieties Grown in the Punjab of Pakistan, 1979–1985.

Note: Average pair-wise coefficients of diversity for varieties grown in each year. Coefficient of diversity = 1–coefficient of parentage.

9.4. ESTIMATING THE EFFECTS OF GENETIC RESOURCES AND DIVERSITY ON YIELD AND YIELD STABILITY

Improvements in the quality of inputs may result in greater crop production. When the crop output has a market value and a supply response to changes in the quantities of inputs used can be observed, it is possible to estimate, using statistical methods, the value of the contribution of inputs that are traded on markets and those that are not—as well as certain attributes of inputs or input quality. This approach is also appropriate for valuing the use of germplasm resources in varieties released by plant breeding programs (Gollin and Evenson, 1998; see other studies included in Evenson, Gollin, and Santaniello, 1998).

Measuring the contribution of an improved input is complicated by the simultaneous use of other inputs, and estimation of a production function through regression analysis is commonly used to assign contributions of different sources to output changes while holding the effects of other inputs constant. Here, the contribution of genetic diversity and genetic resource type has been estimated along with that of conventional inputs using the Cobb–Douglas functional form, which is widely used in partial productivity studies:

$$(9.1) \qquad YLD_{ht} = A \prod_{i=1}^{m} X_{hit}^{\beta_i} \prod_{j=1}^{n} Z_{hjt}^{\beta_j}$$

where YLD_{ht} = metric tons of wheat per hectare for the h–th observation in period t, X_{hit} = the i–th conventional input, including education and weather effects, of the h–th observation in period t, and Z_{hjt} = the j–th genetic resource or diversity input of the h–th observation in period t.

A possible criticism of including the set of genetic resource and diversity variables directly in any production function analysis is that they are not generally thought of as production inputs over which the farmer makes explicit decisions. When a farmer chooses to plant a wheat variety or a combination of wheat varieties based on their observable characteristics, he or she also chooses to use an unobservable set of genetic resources and their attributes. Alternative specifications may be considered, such as entering the genetic resource and diversity measures as intercept shifters in the estimation.

Although it imposes well-known restrictions on production parameters, the Cobb–Douglas functional form is frequently used in partial productivity studies. All functional forms impose some restrictions, and even when enough is known about underlying relationships to specify them adequately, greater flexibility is achieved with losses in degrees of freedom and increased collinearity (Griffin, Montgomery, and Rister, 1987). Here, the purpose of the investigation is to test hypotheses regarding the effects of unobservable genetic characteristics on the mean and variance of output rather than the magnitude of production elasticities.

A yield stability equation has also been estimated to focus more specifically on the effects of genetic resource use and diversity on yield variation. A frequent choice of dependent variable for yield stability studies using panel data is either the coefficient of variation or the Cuddy–Della Valle index. This study uses an alternative method to isolate yield variability in order to preserve limited observations. The dependent variable in the yield stability equation was obtained by removing the trend from mean yield over the seven-year period using a linear regression. With the effect of yield increases removed from the mean, the new dependent variable is calculated by subtracting the mean yield adjusted for trend from each district's observed yield. The regression model was specified as:

$$(9.2) \qquad \left(Y\text{-}D\bar{Y}\right)_{ht} = \beta_0 + \sum_{i=1}^{m} \beta_i X_{iht} + \sum_{j=1}^{n} \beta_j X_{jht} + e$$

where $(Y\text{-}DY)_{ht}$ = absolute value of yield first difference of the h–th observation in period t, X_{iht} = the i–th conventional explanatory input, such as weather and fertilizer effects, of the h–th observation in period t, and Z_{jht} = the j–th genetic resource or diversity input of the h–th observation in period t.

Alternatives to this method for estimating stochastic production technologies include the Just–Pope specification used by Widawsky and Rozelle in chapter 10. Through assuming a beta distribution rather than a normal distribution for crop output, Nelson and Preckel (1989) recovered information about the effects of fertilizer on the skewness as well as the mean and variance of corn yields. Love and Buccola (1991) have proposed a method of jointly estimating production technologies and the risk preferences of farmers, which increases the efficiency and consistency of parameter estimates when input use is endogenous. This fuller, economic decisionmaking model may be better suited for estimating technology and risk parameters for conventional inputs, however, than for estimating the effects on crop production of unobserved genetic resource characteristics of varieties. Farmers choose varieties unaware of the attributes of the genetic resources whose effects we are attempting to measure, unless these are immediately observable or "marked" through their strong linkage to observable plant characteristics (as in the case of the semidwarf wheats and plant height). This implies, among other things, that the supply and demand aspects of the economic decision can be treated as separable.

9.5. ECONOMETRIC RESULTS

9.5.1. Mean Yield
The variables representing conventional inputs, genealogical distance, and the concentration of area among varieties differed significantly at the mean between rainfed and irrigated production environments. Results of a Chow test confirmed that separate regressions should be estimated for irrigated and rainfed production environments. In recognition that omitted variables, other errors, or changing structural circumstances may lead to changing cross-sectional and time-series intercepts, the method used to pool the cross-sectional and time-series data was to include an indicator variable for time in each regression. A test of heterogeneity confirmed that intercepts varied by time period.

Regression results for the effects on yield of conventional inputs and genetic resource and diversity indicators are shown in Table 9.3. Since economic theory predicts positive

Table 9.3. Effects on Yield of Genetic Resource and Diversity Indicators

Explanatory variable	Irrigated	Rainfed
Constant	4.493	−8.154**
	(1.20)	(−2.41)
Log[Rain]	0.027*	−0.167
	(1.910)	(−1.84)
Log[Bullocks]	0.097*	−0.707
	(1.88)	(−2.61)
Log[Fertilizer]	0.232*	0.076*
	(6.029)	(2.03)
Log[Irrigation]	0.558*	0.073*
	(8.55)	(3.266)
Log[Labor]	−0.287	0.391*
	(−2.84)	(4.748)
Log[Literacy]	0.263*	−0.044
	(3.419)	(−0.905)
Log[Tractors]	0.019	−0.052
	(0.51)	(−1.650)
{F}Year[80]	0.167**	0.078**
	(3.73)	(2.260)
{F}Year[81]	0.173**	0.011
	(2.87)	(0.242)
{F}Year[82]	0.204**	0.169**
	(2.72)	(2.264)
{F}Year[83]	0.048	−0.084
	(0.56)	(−1.084)
{F}Year[84]	0.193	−0.214**
	(1.74)	(−2.1)
{F}Year[85]	0.363**	0.013
	(2.92)	(0.116)
Log[Genealogical distance+1]	−0.012	0.168**
	(−0.37)	(2.476)
Log[Age of varieties]	−0.399**	0.056
	(−2.03)	(1.041)
Log[Area concentration]	0.052**	0.028
	(2.12)	(0.376)
Log[Landraces]	−0.599	1.765**
	(−0.44)	(2.131)
Log[Parental combinations]	−0.258	−0.118
	(−0.42)	(1.366)
R^2	0.82	0.98
n	175	28

Note: t-values in parentheses; * indicates significance at 5%, one-tailed test; ** indicates significance at 5%, two-tailed test.

marginal products and production elasticities in the relevant range of production, one-tailed t-tests were conducted on all estimated coefficients for conventional inputs. Coefficients of conventional inputs for irrigated areas are significantly positive except for the labor and tractor use coefficients. This result is surprising because of labor shortages in the favorable environments, although the construction of the labor variable may have actually overstated (or simply mis-measured) the true level of labor use. Also of interest, but without explanation, is the lack of significance of the coefficient for tractor use, given the increased role of tractors in wheat production in irrigated areas.

In the predominantly rainfed areas, labor, irrigation, and fertilizer have significant, positive effects on mean yields. The effects of rainfall, literacy, and use of tractors and bullocks are insignificant. Wheat production in rainfed areas relies primarily on residual moisture from the summer season, and therefore yearly cumulative annual rainfall at one location is likely to be a poor predictor of soil moisture availability in different districts during the dry season when wheat is grown. Labor and mechanization variables may be inversely related, capturing the same effect.

Two-tailed t-tests were conducted on the coefficients of the genetic resource use and diversity variables, since economic theory provides no *a priori* predictions of their signs. In the irrigated areas, only the age of varieties and their area concentration are significantly associated with mean yield. A positive sign for area concentration makes sense, because increased area planted to the highest yielding variety will also increase total yields, and farmers in the irrigated areas tend to grow high-yielding varieties (see Heisey *et al.* 1997: 727). A negative sign on age of varieties is consistent with the notion that low varietal turnover implies the continued use of varieties whose disease resistance is weakened. Furthermore, slow turnover denies producers the benefits of new varieties with greater yield potential.

In rainfed areas, only the coefficients for genealogical distance and landraces are significant. A positive and significant coefficient on genealogical distance suggests that increasing dissimilarity in the group of varieties grown by farmers is associated with enhanced yield. If this finding indeed reflects wheat breeding strategies, it suggests that the incorporation of materials with diverse genetic backgrounds may have contributed to broad adaptation. Materials released during the time period under study were not targeted to rainfed conditions, but several of them, including Lyallpur-73 and Pak-81, were very popular among farmers in both irrigated and rainfed environments.

9.5.2. Yield Stability

Table 9.4 presents the regression results for the yield stability model. Because the trend was removed from the dependent variable in the yield stability model, it is not necessary to include indicator variables for each year in order to pool the cross-sectional, time-series data set. Two-tailed tests were conducted on all coefficients. A positive (negative) sign on regression coefficients implies a decreasing (increasing) effect on yield stability among the districts. Results suggest that those variables that contribute to decreasing yield stability are fertilizer use, age of varieties, and the number of different landraces in pedigrees. All three are significant at the 1% level. The only variable that appears to have a significantly positive effect on yield stability is genealogical distance.

This result supports the hypothesis that a wider genealogical distance among cultivated varieties may be associated with greater aggregate yield stability throughout the districts of Punjab. The effect of fertilizer use is not surprising, since fertilizer use can be either variance-increasing or variance-decreasing, depending on the production circumstances. Older age of varieties is likely to increase aggregate yield fluctuations because obsolete varieties suffer yield losses, although this depends on local disease pressure and weather.

9.6. CONCLUSIONS

Questions concerning the diversity of crop genetic resources have prompted economists to study the relationships between types of genetic resources, genetic diversity, and production outcomes in farmers' fields. Wheat productivity and yield stability are important considerations both to individual farmers and national authorities where wheat is the staple food crop. We have sought to augment the understanding of genetic resource diversity and use by (1) examining patterns of varietal diversity in farmers' fields, both at the regional and district levels; and (2) estimating econometrically the mean yield and stability effects of their diversity.

This study finds evidence of increasing genetic resource use and diversity in the Punjab of Pakistan over the brief time period of the study, as illustrated by the number of distinct landraces and parental combinations appearing in the pedigrees of wheat varieties grown in farmers' fields, as well as by indices of spatial diversity and genealogical distance. Temporal diversity declined during the study period, however. Disaggregated analysis at the district level demonstrates greater variability among indicators than is observable at the provincial level, reflecting differences in production environments and in the availability of suitable varieties. Generally, the most productive irrigated areas show a cyclical pattern of varietal adoption characterized by more rapid

Table 9.4. Effects on Yield Variability of Genetic Resource and Diversity Indicators

Explanatory variables	Regression coefficient
Constant	-3.368***
	(-4.37)
Rain	-0.00005
	(-1.03)
Fertilizer	0.001***
	(3.17)
Genealogical distance	-0.01**
	(-2.52)
Age of varieties	0.06***
	(6.60)
Area concentration of varieties	-0.05
	(-0.64)
Landraces in pedigree	0.078***
	(3.56)
Parental combinations in pedigree	-0.004
	(-1.37)
R^2	0.29
n	203

Note: Dependent variable is yield variability (see text and Table 9.1 for construction); t-values in parentheses; ** indicates significance at 5%, two-tailed test; *** indicates significance at 1%, two-tailed test.

turnover but a higher concentration of area among fewer varieties. In the more marginal rainfed areas there is a higher concentration of area in a single variety and lower varietal turnover.

The contribution of the various types of genetic resources and diversity to wheat production and stability also differs by production environment. In the rainfed areas, genealogical distance and number of landraces in the genetic background of varieties are positively associated with mean yield. In the irrigated areas, only the concentration of area among fewer varieties and the age of varieties has a significant impact on yield. The positive sign on the coefficient of area concentration implies that as more area is planted to a single variety (presumably the highest yielding variety), yields rise. The negative coefficient on age of varieties demonstrates that slow varietal replacement has the effect of depressing yield. When more area is concentrated among fewer varieties, however, spatial diversity decreases and the risk of yield losses from disease increases. An increased rate of varietal replacement in farmers' fields counters the likelihood of an epidemic occurring but requires a highly organized and efficient seed multiplication and distribution system.

The yield stability equation suggests that greater genealogical distance and increased temporal diversity are associated with reduced yield variability among the districts of Punjab from 1979 to 1986. The positive effect of genealogical dissimilarity on yield stability may capture the effects of a broader genetic background on the adaptability of varieties across locations. The negative relationship of variety age and yield stability may reflect the fact that higher rates of varietal turnover provide new sources of disease resistance to farmers, counteracting the destabilizing influence of uneven disease pressure across locations and time.

Acknowledgments

The authors thank Efrén del Toro, previously a statistician with the CIMMYT Wheat Program, for assistance in preparing genealogical data, as well as Derek Byerlee (the World Bank) and Mubarak Ali (Asian Vegetable Research and Development Centre) for assistance in locating and assembling the production data used in this study. This chapter is based on the M.S. thesis of Jason Hartell, and a more detailed account of this research can also be found in a CIMMYT Economics Working Paper (Hartell *et al.*, 1997).

References

Brennan, J. P., and D. Byerlee. 1991. The rate of crop varietal replacement on farms: Measures and empirical results for wheat. *Plant Varieties and Seeds* 4: 99–106.

Cooper, D., R. Vellvé, and H. Hobbelink. 1992. *Growing Diversity: Genetic Resources and Local Food Security*. London: Intermediate Technology Publications, London.

Cox, T. S., J. P. Murphy, and D. M. Rodgers. 1986. Changes in genetic diversity in the red winter wheat regions of the United States. *Proceedings of the National Academy of Science* (US) 83: 5583–5586.

Dalrymple, D. G. 1986. *Development and Spread of High-Yielding Wheat Varieties in Developing Countries*. Washington, DC: US Agency for International Development, Bureau for Science and Technology.

Duvick, D. N. 1984. Genetic diversity in major farm crops on the farm and in reserve. *Economic Botany* 38(2): 161–178.

Evenson, R. E., D. Gollin, and V. Santaniello (eds.). 1998. *Agricultural Values of Plant Genetic Resources*. Wallingford: CAB International.

Frankel, O.H. 1970. The genetic dangers of the Green Revolution. *World Agriculture* 19(3): 9–13.

Gollin, D., and R. E. Evenson. 1998. An application of hedonic pricing methods to value rice genetic resources in India. In R. E. Evenson, D. Gollin, and V. Santaniello (eds.), *Agricultural Values of Plant Genetic Resources*. Wallingford: CAB International.

Griffin, R. C., J. M. Montgomery, and M. Edward Rister. 1987. Selecting functional form in production function analysis. *Western Journal of Agricultural Economics* 12 (2): 216–227.

Hartell, J. G. 1996. *The Contribution of Genetic Resource Diversity: The Case of Wheat Productivity in the Punjab of Pakistan*. M.S. thesis, University of Minnesota, Department of Applied Economics, St. Paul, Minnesota.

Hartell, J., M. Smale, P. W. Heisey, and B. Senauer. 1997. *The Contribution of Genetic Resources and Diversity to Wheat Productivity: A Case from the Punjab of Pakistan*. CIMMYT Economics Working Paper 97-01. Mexico, D.F.: International Maize and Wheat Improvement Center (CIMMYT).

Heisey, P. W., M. Smale, D. Byerlee, and E. Souza. 1997. Wheat rusts and the costs of genetic diversity in the Punjab of Pakistan. *American Journal of Agricultural Economics* 79: 726–727.

Love, H. A., and S. Buccola. 1991. Joint risk preference-technology estimation with a primal system. *American Journal of Agricultural Economics* 73: 765–774.

National Research Council, Committee on Genetic Vulnerability of Major Crops. 1972. *Genetic Vulnerability of Major Crops*. Washington, DC: National Academy of Sciences.

Nelson, C. H., and P. V. Preckel. 1989. The conditional beta distribution as a stochastic production function. *American Journal of Agricultural Economics* 71: 370–378.

Pardey, P. G., J. M. Alston, J. E. Christian, and S. Fan. 1996. *Hidden Harvest: US Benefits from International Research Aid*. Washington, DC: International Food Policy Research Institute (IFPRI).

Souza, E., P. N. Fox, D. Byerlee, and B. Skovmand. 1994. Spring wheat diversity in irrigated areas of two developing countries. *Crop Science* 34: 774–783.

Weitzman, M.L. 1992. On diversity. *Quarterly Journal of Economics* 107: 363–406.

Weitzman, M. L. 1993. What to preserve? An application of diversity theory to crane conservation. *Quarterly Journal of Economics* 108: 1557–183.

10 VARIETAL DIVERSITY AND YIELD VARIABILITY IN CHINESE RICE PRODUCTION

D. Widawsky and S. Rozelle

10.1. INTRODUCTION

Inter-varietal diversity, broadly defined as the extent of dissimilarity among a set of varieties, is increasingly recognized as important to crop production and has commanded growing attention from research in recent years. Farmers and agricultural policymakers may have an interest in varietal diversity because no single variety can completely resist or tolerate all potential stresses, and yield reduction from a particular stress may be lower, on average, when there are more sources of stress tolerance. By providing a broader base of stress tolerance, varietal diversity may also reduce yield variability when pest infestations strike or bad weather occurs.

An important gap in the diversity literature is the lack of research addressing the fundamental question: What *impact* or value is there in maintaining varietal diversity? In industrialized countries, most studies of crop diversity are limited to the genetic analysis of varieties and breeding stock, and they rarely address the impact of diversity. In developing countries, little is even known about crop diversity, and virtually nothing is known about how varietal diversity affects production at the household, regional, or national level. Such a lack of attention is surprising given the large proportion of people that rely on farming and the great importance that developing country governments attach to the food economy.

The goal of this chapter is a better understanding of diversity in the modern rice varieties grown in developing countries, taking eastern China as an example. We begin by exploring several indices of varietal diversity based on numbers of varieties, their genealogies, and planted area. We then use a diversity index constructed from genealogical information about varieties to assess the level of varietal diversity and its impact on the mean and variance of rice production in China. When yield and variance elasticities are estimated with our data set and the diversity index is applied, the resulting estimates suggest that an increase in diversity entails a slight reduction in average yields and a larger reduction in yield variability.

Our study focuses on assessing the impact of diversity at the *regional* level, concentrating on rice-growing areas in Zhejiang and Jiangsu Provinces, which are located in the lower reaches of China's Yangtze River Valley. Eastern China is an excellent site for assessing the regional impacts of varietal diversity in rice production. Yangtze delta rice farmers use high levels of inputs and have a long history of adopting new varieties. China's government has invested substantially in research to develop improved rice varieties, and plant breeders at the national, regional, and local level have developed many new varieties in recent years (Stone, 1988). Regional leaders in this part of China also bear much of the responsibility for maintaining high and stable grain supplies (Rozelle, 1994). At the same time, regional agricultural leaders can influence which rice varieties households select, which rice varieties are available in government-run seed companies, and which varieties are promoted by the public extension system. In our investigation of regional impacts of varietal diversity, we use townships as the unit of analysis.

This chapter is divided into three parts. The first part describes three measures of varietal diversity, which are applied to township data in the second part of the chapter to explore the relationship between varietal diversity and yield variability. The third part of this chapter models this relationship more formally using a stochastic production function. We conclude with a summary and a discussion of the results.

10.2. MEASURING VARIETAL DIVERSITY

10.2.1. Variety Counts

Counting the number of varieties in a given area—varietal richness—is a simple, common method for rapidly assessing diversity in an area. Any index of this type has the drawback that if several varieties are grown on very small plots, and a single variety is grown on the rest of the sown area, the number of varieties may not adequately describe the distribution (abundance) of varieties. One alternative measure of spatial diversity, which captures richness as well as abundance, is a count of the number of varieties accounting for a fixed proportion of total sown area for a given crop (for example, the number of varieties planted on 90% of a region's rice area). However, neither this index nor a count of varieties can distinguish underlying differences among varieties. Two similar varieties could be counted as different varieties when they may actually be closely related (see chapter 2). Similarities among varieties are embodied in their genotypes, and one way to compare genotypes is with pedigree information. Pedigree information is the basis of the other diversity measures used in this study.

10.2.2. Solow/Polasky Diversity

Solow and Polasky (1994) propose an index of species diversity based on the notion of genetic distance. We use their measure to calculate diversity among a set of rice varieties based on genealogical distances derived from the pedigrees of varieties. For a given set of varieties, the index summarizes the genealogical distances between all pairs of varieties while accounting for contemporaneous similarities among all varieties. The Solow/Polasky index has at least three desirable and fundamental properties. First, it is monotonic in varieties, implying that diversity should not be decreased by the addition of a variety. Second, it accounts for twinning, meaning that diversity does not increase when we add a variety identical to one already included in the set. Third, the Solow/Polasky measure is monotonic in distance; diversity should not be decreased by an unambiguous increase in the distance between species.

The Solow/Polasky diversity measure can be calculated as:

(10.1) $V(S) = e' \, F^{-1} \, e$

(10.2) $f_{jk} = \mathrm{Corr}\big(I_j I_k\big) = f(d_{j,k})$

where diversity, $V(S)$, is a function of F, a matrix of pair-wise distances, and e, a vector of 1s (equation 10.1). Larger values of $V(S)$ implies greater diversity. The F-matrix is made up of pair-wise coefficients (f_{jk}), contained in the interval $[0,1]$, that represent the "correlation" between two varieties (equation 10.2). If distance measures are not contained in the $[0,1]$ interval, one may employ a cumulative distribution function that is a function of the pair-wise distance. The elements of the correlation matrix, F, can be interpreted as the probability that two varieties share the same characteristics, which depends on the dissimilarity between them, $d_{j,k}$. Therefore, to use a Solow/Polasky index, it is necessary to obtain a pair-wise varietal distance measure that represents the probability that two varieties exhibit the same characteristics. Such a measure can be obtained from coefficients of parentage.

We use coefficients of parentage (COPs) as the f_{jk} in equation 10.1 to produce a Solow/Polasky index of diversity for sets of varieties in this study (with appropriate data, one could also use pair-wise distance measures based on molecular or morphological characteristics). The COP is a pair-wise comparison that estimates the probability that a random allele take from the same locus in a variety X is identical, by descent, to a random allele take from the same locus in variety Y. Values range from 0 to 1, with higher values indicating greater similarity in ancestry (for more information on COP, see chapter 2). For each pair of varieties within each township, for each year, the COP was calculated using detailed pedigree data and a modification of a computational program developed by Sneller (1994).

10.2.3. Area-weighted Diversity

While the Solow/Polasky measure summarizes differences among a set of sown varieties, it does not account for the abundance or spatial distribution of varieties in any given area. We propose a modification of the Solow/Polasky index to account for the share of total area planted to each variety based on an experimental method suggested by Polasky (personal communication). If s_i is the proportion of sown area accounted for by variety i, and there are n varieties in a township, then an area-weighted distance ($d_{i,j}^{\omega}$) reflects the importance of abundance and is given by:

$$(10.3) \qquad d_{i,j}^{\omega} = \frac{d_{i,j}}{\left[\frac{s_i s_j}{\left(\frac{1}{nxn}\right)}\right]}$$

where d_{ij} are inter-varietal distances based on varietal dissimilarities. When s_i is uniform and all varieties occupy the same proportion of sown area, weights are all one, and $d_{i,j}^{\omega}$ equals d_{ij}. When two varieties' proportion of sown area is greater than $(1/n)$, then pair-wise distance is decreased by abundance, implying that diversity decreases. Such a measure corrects for instances in which large genetic distances suggest substantial diversity, but concentrated sown areas suggest limited diversity.

10.3. EVIDENCE OF VARIETAL DIVERSITY AND YIELD VARIABILITY IN EASTERN CHINA

10.3.1. Data
Zhejiang and Jiangsu Provinces in eastern China are located in China's highly productive rice-growing coastal areas and share similar climatic and agronomic environments. Eight townships in eight counties distributed throughout these provinces were chosen as study sites and surveyed by the authors for each of the years 1984–1991. The regional variations may be helpful in characterizing the relationship between diversity and production and are examined below.

The township was chosen as the unit of observation for several reasons. The centrally planned agricultural system is organized (in descending level of aggregation) into national, provincial, prefectural, county, township, village, and household units. Detailed agricultural data have been collected for several decades at all levels, but the lowest level of reliable data is the township, for which information is available on input use, area planted to different rice varieties, and pest infestation levels. Workers trained at agricultural colleges collect production data throughout the year. The years 1984–1991 were chosen because they represent a recent period relatively free of economic upheaval. A more detailed description of economic and agronomic conditions in the study area is given in Widawsky (1996). Despite the similarities in yields, input use and varietal adoption patterns vary throughout the study areas.

10.3.2. Variety Counts
A count of varieties grown in the study sites suggests that rice diversity in Zhejiang exceeds that in Jiangsu (Figure 10.1). In each Zhejiang sample township, on average, 90% of the township area planted to rice was occupied by a minimum of five to eight varieties in a given year. In contrast, only three varieties represented 90% of the rice area in all but one of the Jiangsu counties. The Jiangsu county with the highest level of count-based diversity still had less diversity than the Zhejiang county with the least diversity.

Greater adoption of hybrid rice varieties in Jiangsu may partially account for differences in this simple measure of diversity (Widawsky, 1996). Adoption of hybrid rice varieties by Jiangsu producers averages more than 50% of sown area, far higher than in Zhejiang. Hybrid rice requires specialized restorer lines to propagate male sterile lines. As a result, there is less breeding material for hybrid rice varieties and fewer cultivars are available for any given location. In areas where conventional rice varieties are grown, farmers have a greater number of varieties to choose from, and according to the index used here, these areas appear to have greater diversity in terms of varietal richness. A comprehensive treatment of differential rates of adoption for conventional versus hybrid varieties in China is given in Huang and Rozelle (1996).

Although average rice yields are higher in the less diverse Jiangsu study sites (6.7 versus 5.8 t/ha in Zhejiang), these locations experience greater yield variability. In every Jiangsu study site, the average coefficient of variation of rice yields between 1984 and 1991 was greater than the coefficient of variation in three out of four Zhejiang counties. The difference is substantial in two of the Jiangsu counties, Gaoyou and Jinjiang.

Yield variability also appears to be correlated with varietal abundance (Figure 10.2). As the number of varieties in an area increases, the coefficient of variation markedly declines (correlation = –0.43). Agricultural officials in the study area indicated that with fewer replacement varieties, a given hybrid variety tends to be grown for more consecutive years. Since pest resistance breaks down over time in the rice varieties grown in eastern China (Widawsky, 1996), longer growing cycles leave high-yielding hybrid varieties more susceptible to some types of pests and hence more vulnerable to yield fluctuations.

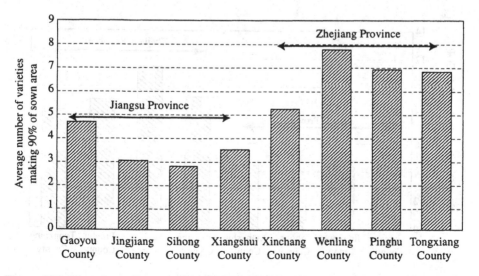

Figure 10.1. Diversity in Eastern China (1984–1991), Based on Variety Counts.

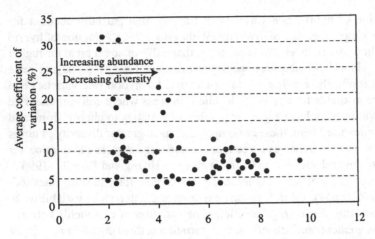

Figure 10.2. Yield Variability and Variety Counts in Eastern China, 1984–1991.

10.3.3. Solow/Polasky Diversity Indices and Yield Variability

The Solow/Polasky index reveals a pattern of diversity similar to that found with the measure based on varietal counts (Figure 10.3a). This result supports the conjecture that as the number of rice varieties increases, so does latent diversity, as measured by genealogical distances. Such a pattern need not necessarily occur. If (as some local observers and government documents have warned) cosmetic breeding of conventional or hybrid varieties were occurring, the genealogical distance among varieties could be small. Diversity could diminish or stagnate even as the number of varieties rises.

Weighting the Solow/Polasky index by the proportion of area sown to each variety diminishes the differences among provinces in the level of rice diversity (Figure 10.3b). One factor that could account for such a pattern is that producers in Zhejiang have a

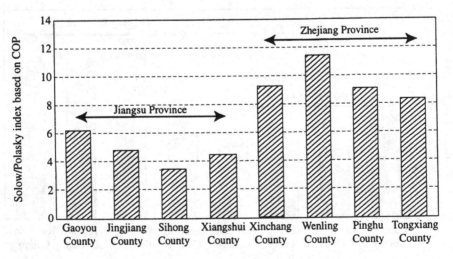

Figure 10.3a. Rice Diversity in Eastern China, 1984–1991.

propensity to plant small fractions of their areas to specialty varieties such as glutinous or high-quality cultivars. They are also more likely to field-test one or more of the many new conventional varieties that are released each year. In such cases, the small proportion of area planted to specialty varieties generates extremely large weights for the pair-wise distances in the area-weighted Solow/Polasky index, and substantially diminishes the differences in diversity between Zhejiang and Jiangsu townships. Field visits to rural Zhejiang did reveal that farmers were interested in planting small plots for special household needs, such as glutinous rice for festivals. While some farmers in Jiangsu also grew specialty crops, the practice was much less apparent at the township level than for Zhejiang.

Overall, the level of rice diversity is greatly reduced when the Solow/Polasky index is weighted by area shares of varieties (Figure 10.3b). While the area-weighted measure penalizes varieties that contribute only small fractions to total area, it is somewhat surprising that diversity is so low. In most areas, apparently, a small number of varieties dominates the mosaic of local varieties. Such a result, however, is less surprising given the recent changes in China's rural areas that encourage specialization and increased efficiency (Sicular, 1995; Rozelle, Pray, and Huang, 1997).

Finally, the inverse relationship between diversity—as measured by the unweighted Solow/Polasky index—and yield variability is similar, but slightly weaker, than that found for counts of varieties (Figure 10.4a, correlation = –0.31). There is no apparent relationship, however, between the weighted Solow/Polasky index and yield variability (Figure 10.4b, correlation = 0.07).

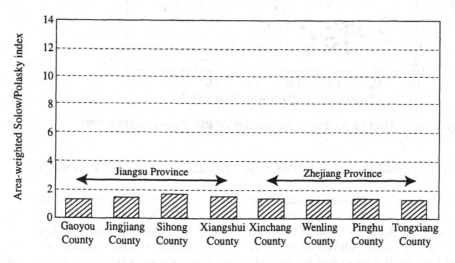

Figure 10.3b. Area-weighted Rice Diversity in Eastern China, 1984–1991.

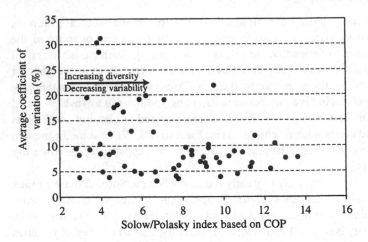

Figure 10.4a. Yield Variability vs. Rice Diversity, 1984–1991.

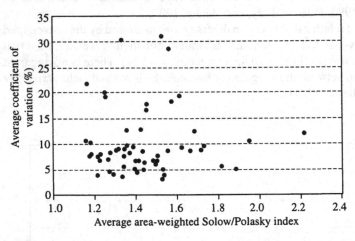

Figure 10.4b. Yield Variability vs. Area-weighted Rice Diversity, 1984–1991.

10.4. ESTIMATING THE IMPACT OF VARIETAL DIVERSITY IN RICE

At least one of the diversity measures explored in the last section suggests a relationship between diversity and yield variability in eastern China: high levels of diversity may be associated with low levels of yield variability. This relationship is important because individuals at various levels of the food production and distribution process value low yield variability. National leaders in China have always been concerned about minimizing food production variability and its destabilizing effects. Performance evaluations of local government leaders depend on the achievement of high, stable yields (Rozelle, 1994). In the past, local leaders could allocate low-cost inputs to guarantee high yields. Government support for farm input subsidies has been reduced, however, and regional rice production increasingly depends on other mechanisms to meet production targets. If higher varietal diversity corresponds to lower yield variability, the implication is that promoting diversity will increase yield stability. Since government officials still influence the availability and adoption of rice varieties (Rozelle, 1994), they are well positioned to promote varietal diversity. Of course, leaders also worry about yield levels, and if diversity entails a tradeoff between increasing yield stability and decreasing average yields, it is critical to develop some sense of the size of this tradeoff. The rest of this chapter examines this issue by testing the impact of varietal diversity on regional rice production, specifically mean yields and yield variability.

10.4.1. A Stochastic Production Model

The impact of varietal diversity on rice yields can be assessed using any one of a number of approaches developed to analyze stochastic production functions. We used the Just and Pope (1979) method:

(10.4a) $Y = f(X,\beta) + u$
(10.4b) $u = h(X,\phi)\varepsilon$

where Y is output, X is a vector of inputs, β is a matrix of parameters of the production function (equation 10.4a), u is an error term, ϕ are parameters of the variance function (equation 10.4b), and ε is an independent and identically distributed random term. In actual practice, after estimating β, the parameters of the mean production function, one can calculate production elasticities. Using the estimates of β, one can then calculate the squared residuals of equation 10.4a as an estimate of variance. Regressing the estimates of variance on a set of explanatory variables leads to flexible estimates of elasticities of variance for the individual explanatory factors (equation 10.4b).

The stochastic production model has several characteristics that are helpful in answering the questions we have posed. Unlike in some single equation production functions, here inputs are not assumed to have the same influence on both average yield and yield variability; inputs can increase yields while decreasing variability. Stochastic production analyses can measure how varietal diversity independently affects mean yields and yield variability.

10.4.2. Specification of a Stochastic Production Function for Eastern China

Average rice production mean and yield variability are empirically modeled in two stages.[1] Because pesticide use in the yield equation may be endogenous, pesticide is the dependent variable in the first stage:

(10.5) $z_p = a_0 + \sum_{i=1}^{r} a_i A_i + \sum_{j=1}^{r} d_j P_j + \sum_{h=1}^{7} b_h L_h + cQ + v$ $p = 1,2$ (insecticide, fungicide)

Insecticide and fungicide are scaled by their geometric mean and represented in log form (z_p). In equation 10.5, pesticide use is a function of area infested by each pest (A, given as a percentage), severity of infestation (P, as incidents of multiple pesticide treatments), location dummies to account for fixed effects (L), time as a proxy for technical change, and grain quota obligations (Q). The error term is assumed to be normally distributed with zero mean and variance, w^2. The model was estimated by ordinary least squares and used to generate predicted values of pesticide for use in the second stage.

The second stage generates estimates of equations 10.4a and 10.4b and is empirically specified as:

$$(10.6) \quad Y = e^{\alpha_0} \left(\prod_{i=1}^{n} X_i^{\alpha_0} \right) \left(\prod_{j=1}^{4} Z_j^{\beta_j} \right) \left(Z_1^{\beta_5 \ln Z_3} \right) \left(Z_2^{\beta_6 \ln Z_4} \right) \left(D^{\lambda} \right) \left(\prod_{k=1}^{m} e^{\lambda_k S_k} \right) \left(\prod_{h=1}^{7} e^{\delta_n L_n} \right) + u$$

$$(10.7) \quad u^2 = \left[h(X,Z,\phi,\theta)\varepsilon \right]^2 = e^{\phi_0} \left(\prod_{j=1}^{4} X^{\phi_j} \right) \left(Z_1^{\phi_5 \ln D} \right) \left(Z_3^{\phi_6 \ln D} \right) \left(D^{\phi_7} \right) \left(\prod_{i=1}^{3} X_i^{\theta_i} \right) \left(e^{\theta_i S_i} \right) e^{v}$$

Rice yield (equation 10.5) is a function of "standard" inputs (X), pesticide use and host-plant resistance as pest control inputs (Z), other shifter variables (S), a set of county dummy variables (L), and diversity (D). The vector of pest control inputs (Z) includes insecticide (Z_1), fungicide (Z_2), host-plant resistance to insects (Z_3), and host-plant resistance to diseases (Z_4). The vector of shifter variables (S) includes incidence of double cropping, climatic disaster, and time as a proxy for technological development. Locational dummies (L) are given for seven of the eight counties. We estimated the model twice, with the unweighted and area-weighted Solow/Polasky indices. Variables in equation 10.6 were scaled by geometric means (Boisvert, 1982). The yield equation is specified as a generalized Cobb–Douglas, and was estimated with non-linear least squares because the error term (u) is additive and the equation is not linear in parameters.

Variance also is modeled as a Cobb–Douglas approximation of a general functional form (equation 10.7). The variance of yield is explained by pest control inputs (Z), nitrogen, labor, mechanized input, double cropping, and varietal diversity (D). The variables in the variance equation are specified in log form and the equation is estimated using ordinary least squares. Because equation 10.6, the yield equation, uses data scaled by geometric means, the estimated squared residuals, $\hat{u}^2$, are multiplied by the square of the geometric mean of output to generate the approximations of estimates of output variance that are used in equation 10.7. This is because equation 10.6 regresses scaled yield $(Y/\dot{Y})$ on explanatory variables, where $\dot{Y}$ is the geometric mean of output. Since variance of actual output is the item of interest, it is necessary to multiply the estimated variance from the output equation by $\dot{Y}^2$, because $\text{var}(Y/\dot{Y}) = \text{var}(Y)/(\dot{Y}^2)$. Given the functional form, yield and variance elasticities are parameters and can be read directly from the regression output.

10.4.3. Data

To estimate equations 10.5–10.7, we used data collected from 64 townships in eight counties in Zhejiang and Jiangsu Provinces. Accountants and agronomic technicians provided township data, including information on standard production inputs (nitrogen fertilizer, phosphorous fertilizer, agricultural labor, machinery) and other information such as the incidence of double cropping and natural disasters in the township. In addition, we collected information to construct several measures of pesticide use and host-plant pest resistance, factors that may complement or substitute for the impact of varietal diversity on rice yields in eastern China. Detailed information on sown varieties, their pedigrees, and area planted to each variety were used to construct the indices of varietal diversity, as described in the previous section. A summary of the data appears in Widawsky (1996).

10.5. RESULTS

The model is estimated as indicated above and regression summaries of the mean and variance equations are given in Tables 10.1 and 10.2.[2] A separate model was estimated for the unweighted and area-weighted Solow/Polasky indices.

The general performance of the mean yield model was good, with an asymptotic R^2-value of 0.93 (Table 10.1). Conditions in eastern China emphasize high yields, and the high rate at which traditional inputs are used leads to low marginal productivities. The estimated production elasticities of fertilizer inputs are generally quite low and not significantly different from zero. The production elasticity of labor was 0.04 in both models, while the production elasticity of machinery ranged from 0.014 to 0.021. The total elasticity for current inputs is low and similar to those from studies of Chinese agriculture by Putterman (1993), Kim (1990), and Weimer (1990), which used production team data from the early 1980s and estimated production elasticities from current inputs between 0.05 and 0.11. The positive coefficient on the time variable supports the notion that technological progress took place during the survey period and is consistent with other studies of rice yields in China (Huang and Rozelle, 1996). Double cropping was associated with lower yields than single cropping, a result of the shorter growing periods necessary for planting two rice crops per year. The coefficient on disasters was predictably negative in both cases.

Table 10.1. Estimation Results of Stochastic Production Function: Mean Function

Variable	Unweighted diversity	Asymptotic t-statistic	Area-weighted diversity	Asymptotic t-statistic
Intercept	.963	56.64	.977	54.27
Insect resistance	.021	0.87	.039	1.56
Disease resistance	−.030	−1.11	−.046	−1.70
Insecticide	.020	1.43	.018	1.28
Fungicide	−.025	−2.50	−.033	−3.30
Insecticide/insect resistance interaction	.012	0.32	.018	0.45
Fungicide/disease resistance interaction	−.030	−1.03	−.056	−1.87
Diversity	−.049	−4.90	−.009	−0.06
Inorganic nitrogen	.016	0.94	.017	1.00
Inorganic phosphorous	.000	0.10	.001	0.17
Labor	.040	3.07	.040	2.86
Mechanized inputs	.014	1.55	.021	2.33
Time	.004	2.00	.005	2.50
Double cropping	−.298	−10.27	−.334	−11.51
Disaster	−.163	−7.41	−.162	−7.04
Xinchang County	.219	5.76	.219	5.61
Wenling County	.289	8.26	.278	7.51
Pinghu County	.339	10.27	.333	9.79
Tongxiang County	.264	7.76	.271	7.74
Gaoyou County	.244	10.61	.219	9.52
Jingjiang County	.200	8.00	.222	8.88
Sihong County	.011	0.52	.017	0.77
Asymptotic R^2	0.933		0.933	

Table 10.2. Estimation Results of Stochastic Production Function: Variance Function

Variable	Unweighted diversity	Asymptotic t-statistic	Area-weighted diversity	Asymptotic t-statistic
Intercept	.369	1.47	.373	1.59
Diversity	−.418	−1.59	−.149	−0.39
Insect resistance	−.698	−1.31	−.429	−0.79
Disease resistance	.703	1.12	.484	0.81
Insecticide	.322	1.12	.210	0.72
Fungicide	.055	0.24	.136	0.58
Diversity/insecticide interaction	.919	2.37	−.835	−1.00
Diversity/fungicide interaction	.019	0.07	−.436	−0.95
Inorganic nitrogen	−.105	0.33	−.415	−1.29
Labor	.556	2.02	.363	1.33
Mechanized inputs	.062	0.44	.004	0.03
Double cropping	−.693	1.68	−.728	−1.93
F-statistic	2.79		1.413	
prob(F>F-statistic)	(0.002)		(0.164)	

The general performance of the yield variance model was not as good as that of the mean model. The F-statistic for the model using an unweighted Solow/Polasky index was significant at the 0.2% level, suggesting that the estimated parameters are jointly different from zero. Labor was found to be variance-increasing, while double cropping reduced yield variability. The overall performance of the variance model using an area-weighted Solow/Polasky index was much poorer, with an F-statistic significant only at the 16% level. In addition, asymptotic t-tests suggest that most of the parameters are not statistically different from zero.

Tests on the diversity indices, while mixed, suggest that varietal diversity may reduce yields slightly, while also reducing yield variability. Diversity in the mean yield equation estimated with the unweighted Solow/Polasky index was significant with a production elasticity of −0.049, suggesting that higher levels of diversity depress yields (Table 10.1). At the same time, the estimate of the elasticity of variance (−0.418) suggests that diversity can have a larger impact in reducing yield variability, although the parameter estimate is statistically weak. The significant coefficient on the interaction term between diversity and insecticide suggests that diversity dampens the yield-destabilizing effects of insecticide. In comparison, while the model using an area-weighted Solow/Polasky index generates production and variance elasticities with the same sign as the unweighted model, t-statistics are so low that results are inconclusive. The apparent lack of relationship between the area-weighted Solow/Polasky index and the variance is supported by the descriptive data in Figure 10.4b.

The estimates also suggest interesting relationships between pest control inputs and yields, although the results are not definitive. Conventional wisdom holds that pesticides are not only yield-increasing but variance-reducing technologies. The insurance value of yield stability is thought to be one reason why producers use pesticides, even when pesticides have low production elasticities. In our results, elasticity estimates suggest that insecticides increase yields slightly but also increase yield variability. If true, this would suggest that there is little insurance value from insecticides in Chinese rice production. The low t-statistics of our parameter estimates suggest that further study is warranted. At the same time, fungicides appear to depress mean yields significantly and have no discernible effect on yield variability. The overall results with respect to host-plant resistance to insects and disease were similarly mixed.

10.6. SUMMARY AND CONCLUDING COMMENTS

With decreasing government involvement in Chinese agricultural production and grain markets, information on rice yield variability could be important to local leaders, particularly since such leaders work to insure against regional production risk with a shrinking set of tools at their disposal. In addition, the risk-reducing benefit of varietal diversity is likely to be most pronounced at the township or regional level where diversity is greater. While individual households in high-input cropping systems may grow as many as several rice varieties, low per capita land availability in eastern China probably limits the reduction in yield variability a given farmer might experience in diversifying his rice varieties.

In this study, we have tried to measure the impact of the diversity of crop genetic resources on mean yields and yield variability by including an indicator of diversity in a production model. To incorporate concepts related to genetic diversity into economic analyses, it is first necessary to measure diversity using established criteria and to understand the scope and limitations of the indicators one might employ. Counting varieties or species can be a robust measure of diversity in some cases. But such a measure may also fall short in describing the underlying dissimilarity among varieties when measures of dissimilarity are not proportional to the number of sown varieties. At the same time, diversity indices based on pair-wise distance measures of varietal similarity may fail to describe the distribution of varieties in an agricultural landscape.

In analyzing species diversity, Solow and Polasky developed an index based on pair-wise distances among members of a set. We have applied this index to sets of rice varieties grown in eastern China at different locations and times, using coefficients of parentage as the pair-wise measures of distance. We also tried to modify the index to account for the distribution of varieties using the proportions of sown area in each variety. With these indices, we then used one of several available methods to estimate a stochastic production model and explored the role of varietal diversity in rice yields and yield variability in China.

Results were mixed, although they suggest that, by some indicators, greater rice diversity may decrease yields slightly, while reducing its variability. Using different types of diversity indices in similar economic analyses—or even using Solow/Polasky indices based on other distance measures such as plant morphology—could prove useful in increasing our understanding of the role of plant genetic resources in crop production. There is still much work to be done in developing richer economic models to analyze the risk implications of regional crop diversity measured over time and across different locations.

It is a fact in China—and in many other developing countries—that market liberalization in the agricultural sector is creating conditions where risk management will more often become the responsibility of individual farmers and local agricultural officials. As farmers make production decisions under these new economic conditions, understanding the implications of these decisions becomes an important focus of economic analysis. It is also a fact that the diversity of crop genetic resources is a topic of increasing importance to policymaking in many fora. By establishing links between varietal diversity and agricultural production, economists can and should make important contributions to the formation and implementation of related policy.

Acknowledgments

This research was made possible by grants from the Rockefeller Foundation, International Development Research Centre (IDRC), the Morrison Institute, and Resources for the Future.

Notes

1 Details on data and other estimations are found in Widawsky (1996), where the econometric model was used in a study of the impact of pest control and varietal diversity in eastern China.
2 Regression results for pesticide prediction equations are not included here because of space limitiations and because pesticide endogeneity is more germane to pesticide productivity, whereas this chapter focuses on varietal diversity. A complete description of the pesticide prediction estimates, including tests showing that pesticides may be endogenous, can be found in Widawsky (1996).

References

Boisvert, R. N. 1982. *The Translog Production Function: Its Properties, Its Several Interpretations and Estimation Problems*. Research Bulletin. A.E.Res.82–88. Ithaca: Department of Agricultural Economics, Cornell University.

Huang, J., and S. Rozelle. 1996. Technological change: Re-discovering the engine of growth in China's rural economy. *Journal of Development Economics* 49: 337–369.

Just, R. E., and R. D. Pope. 1979. Production function estimation and related risk considerations. *American Journal of Agricultural Economics* 61: 276–284.

Kim, S. J. 1990. Productivity effects of economic reforms in China's agriculture. Ph.D. thesis, Brown University, Providence, Rhode Island.

Putterman, L. 1993. *Continuity and Change in China's Rural Development: Collective and Reform Eras in Perspective*. New York: Oxford University Press.

Rozelle, S. 1994. Decision-making in China's rural economy: The linkages between village leaders and farm households. *China Quarterly* 137: 99–124.

Rozelle, S., C. Pray, and J. Huang. 1997. Agricultural research policy in China: Testing the limits of commercialization-led reforms. *Comparative Economic Studies* 39: 37–71.

Sicular, T. 1995. Redefining state, plan, and market: China's reforms in agricultural commerce. *China Quarterly* 144: 1020–1046.

Sneller, C. H. 1994. Documentation for SAS coefficient of parentage programs. Fayetteville: Department of Agronomy, University of Arkansas.

Solow, A., and S. Polasky. 1994. *Measuring Biological Diversity*. Working Paper, Woods Hole Oceanographic Institution. Woods Hole: Woods Hole Oceanographic Institution.

Stone, B. 1988. Developments in agricultural technology. *The China Quarterly* 116: Special Issue on Food and Agriculture in China During the Post-Mao Period (December).

Weimer, C. 1990. Reform and the constraints on rural industrialization in China. Mimeo. Honolulu: Department of Economics, University of Hawaii.

Widawsky, D. 1996. Rice yields, production variability, and the war against pests: An empirical investigation of pesticides, host-plant resistance, and varietal diversity in eastern China. Ph.D. thesis, Food Research Institute, Stanford University, Palo Alto, California.

11 THE ECONOMIC IMPACT OF DIVERSIFYING THE GENETIC RESISTANCE TO LEAF RUST DISEASE IN MODERN BREAD WHEATS

M. Smale and R. P. Singh

11.1. INTRODUCTION

The economic importance of the rust diseases of wheat is without question. The rusts are major historical diseases of wheat and developing genetic resistance to them has been an objective of plant breeding programs since the turn of this century (e.g., Howard and Howard, 1909; Lupton, 1987; Macindoe and Brown, 1968).[1] Periodic rust epidemics were common in most decades of this century, although genetic manipulation of resistance genes over the last 40 years has resulted in more stable patterns of resistance (Singh and Dubin, 1997). Genetic resistance, rather than use of fungicides, remains the principal means of controlling the wheat rusts. This is especially true in wheat-producing countries of the developing world, where the costs of controlling disease outbreaks are relatively high. In many developing countries, procuring and distributing the large quantities of chemicals needed to combat unanticipated epidemics is not always feasible.

Breeding for genetic resistance to rust diseases in wheat is an example of research to maintain crop productivity. Productivity enhancement is measured in terms of positive yield gains associated with research investment; productivity maintenance is estimated in terms of yield losses that would have occurred in the absence of the research investment. Researchers have long argued the importance of maintenance research (Plucknett and Smith, 1986), but there are relatively few economic analyses of maintenance research and, in particular, breeding for genetic resistance to diseases (Brennan, Murray, and Ballantyne, 1994; Collins, 1995; Evenson, 1998; Heim and Blakeslee, 1998; Morris, Dubin, and Pokhrel, 1994; Priestley and Bayles, 1988).

The principal challenge in estimating the benefits to research on genetic resistance to rust is given by the simple fact that rust pathogens evolve: resistance conferred by any given gene or gene combination is usually temporary. The longevity of any new source of resistance is not known when a variety carrying it is released. If detailed, historical farm-level data were available on annual losses from rust over the extensive areas of the wheat-producing world, the benefits could be estimated through tabulation. In the absence of these data, however, benefits must be estimated based on expected losses and predictions of the longevity of resistance.

The study summarized in this chapter represents a special case in estimating the benefits of productivity maintenance research. The objective of the study was to estimate the net benefits from research at CIMMYT to diversify the genetic basis of resistance to leaf rust by predicting the losses that were avoided by switching from a breeding strategy based on race-specific (more simply, "specific") resistance to one based on race-non-specific (or "non-specific") resistance. Through a combination of genetic information, trial data, and data on varietal distributions by area in the Yaqui Valley of Mexico, we develop parameters for simulating net benefits.

11.2. CIMMYT'S STRATEGY FOR INCORPORATING GENETIC RESISTANCE TO LEAF RUST

Since the United States corn leaf blight of 1970, public concern has focused on the potential for plant disease epidemics caused by uniformity in the genetic base of resistance. Wheat breeders and pathologists have known for some time that specific resistance in such diseases as the rusts contributes to a "boom-bust" cycle of resistance and vulnerability, because the pathogen is able to mutate rapidly and form new races (Vanderplank, 1963). The dominant selection methodology used over the past 25 years by CIMMYT's wheat breeding program is based on the concept of non-specific resistance, as defined theoretically by Vanderplank and applied to rust resistance by Caldwell (1968) (see Rajaram, Singh, and Torres, 1988).[2]

CIMMYT's strategy emphasizes the accumulation in varieties of multiple genes conferring partial, race-non-specific resistance. Genes conferring race-specific resistance tend to produce resistant reactions, but their effects are overcome in a relatively short time. By contrast, genes conferring race-non-specific resistance have partial and additive effects, and although the response to infection is essentially susceptible, the rate of disease progress is slowed. Further, non-specific resistance is more likely to endure for many cropping seasons.

Research at CIMMYT has indicated that over the past few decades the impact of breeding for genetic resistance to diseases has generated a large proportion of the total economic return to investment in international wheat research (Byerlee and Traxler, 1995; Sayre et al., 1998). Through CIMMYT's collaboration with national wheat research programs in developing countries, the breadth of the international flow of germplasm containing resistance to rust diseases is likely to have been great. In 1990, CIMMYT and CIMMYT-derived materials were grown on an estimated 73% of the area planted to spring bread wheats in the developing world (Byerlee and Moya, 1993). A recent survey of wheat breeders in developing countries indicates that among the types of materials they use in crossing (including their own advanced lines, advanced lines borrowed from other countries, wild relatives, and landraces), materials from CIMMYT International Nurseries are the most frequently crossed in pursuit of disease resistance goals (Rejesus, Smale, and van Ginkel, 1997).

Several of the major wheat varieties grown in the developing world today, and most of CIMMYT's bread wheat germplasm, contain in their pedigrees the ancestral source of the gene combinations for stem and leaf rust resistance that are believed to confer resistance of a durable nature. For leaf rust, non-specific resistance appears to be based on the additive interaction of the partially effective gene $Lr34$ and two or three other, unnamed, genes (Rajaram, Singh, and van Ginkel, 1997). The ancestral sources of this and other partially effective but longer-lasting genes are believed to be Alfredo Chaves, a Brazilian landrace, and Americano 44D, a Uruguayan variety of unknown origin (Roelfs, 1988).

Table 11.1 shows the extent of cultivation of wheats with these ancestors in their pedigrees. Among the bread wheats grown in developing countries in 1990, a large percentage had either Alfredo Chaves or Americano 44D in their pedigrees. Except in

Table 11.1. Presence of Known Sources of Durable Resistance to Leaf Rust in Pedigrees of Modern Wheats Grown in the Developing World, 1990

Region	Known ancestral source of durable resistance to leaf rust	Percentage of cultivars with source	Percentage of cultivars of CIMMYT origin with source[a]	Percentage of wheat area in cultivars with source
Sub-Saharan Africa	Alfredo Chaves	92	85	97
	Americano 44D	76	89	72
South Asia	Alfredo Chaves	84	92	80
	Americano 44D	46	91	51
West Asia	Alfredo Chaves	80	72	72
	Americano 44D	51	61	49
North Africa	Alfredo Chaves	70	100	71
	Americano 44D	30	100	37
Mexico/Guatemala	Alfredo Chaves	100	100	100
	Americano 44D	93	100	80
Andean Region	Alfredo Chaves	85	83	93
	Americano 44D	55	100	56
Southern Cone, Latin America	Alfredo Chaves	83	82	93
	Americano 44D	67	81	81
Developing world	Alfredo Chaves	85	87	81
	Americano 44D	59	88	55

a Includes only 350 cultivars for which pedigrees are known. CIMMYT "origin" defined as direct parents or grandparents, and thus understates the role of CIMMYT ancestry. The pedigrees of some cultivars contain both ancestors.

North Africa, the percentage of bread wheats with these ancestors was at least as high outside the known region of origin (the Southern Cone of Latin America) as in it. Most of these wheats had at least one parent or grandparent of CIMMYT origin.

We raise these points only to suggest the breadth of the international flow of germplasm. The presence of the source of the *Lr34* complex in the ancestry of a variety does not ensure that the variety contains the gene. Even if the gene is present, interactions with other genes and the environment are important in determining the variety's resistance when challenged by pathogens in farmers' fields. CIMMYT wheat breeders use "bridging crosses" to introduce resistance into their lines, and in the case of the resistance conferred originally by these landrace ancestors, the bridging cross was initially Frontana/Kenya 324//Newthatch and, subsequently, Kalyansona/Bluebird.

11.3. ESTIMATING BENEFITS OF DIVERSIFYING GENETIC RESISTANCE TO LEAF RUST

Our approach in this study was to estimate the benefits of non-specific resistance as the value of the yield losses that farmers in Mexico's Yaqui Valley would have incurred if breeders had employed a strategy for specific resistance rather than non-specific resistance. "Diversity" refers to the presence in the variety of more than one gene, each with a partial effect, as compared to the presence of a single gene with a major effect. The number of genes may still be relatively few.

The magnitude of yield savings associated with non-specific resistance is based on the differences in yield losses that occur for the two types of resistance as a function of the field life of the variety. With specific resistance, yield losses from disease are assumed to be negligible until the pathogen evolves to overcome the resistance. When this occurs, yield losses are large. With non-specific resistance based on partially effective genes, some yield losses from disease may occur even just after the variety is released. Even so, the path of deterioration in resistance, if deterioration occurs, may be more gradual and may not cause "large" losses for many years.

The longevity of resistance varies by the variety and the environment in which it is grown, including farmers' management practices, the spatial distribution of varieties, other crops in the farming system, and weather patterns. In Kilpatrick's 1975 survey of wheat-producing countries, the expected longevity of a variety with specific resistance was about five years, although few data points were reported.

The benefits associated with growing a variety with non-specific resistance occur only when the farmer grows wheat varieties for a longer period than the specific resistance is effective. Farmers may replace their wheat varieties infrequently because of their own preferences for yield or other characteristics, or because of the characteristics of the seed system that constrains their choices (see chapter 14). Patterns of varietal turnover have been reported for the developed and developing world in Brennan and Byerlee (1991) and its determinants analyzed by Heisey and Brennan (1991). For many regions of the developing world the average age of wheat varieties, weighted by the area planted to them, is over ten years (see summary in Smale, 1996).

The economic impact of the resistance breeding strategy was estimated in four steps. In the first step, we estimated with regression analysis the effects on disease losses of non-specific gene complexes, using knowledge of the genetic basis of resistance and experimental data for CIMMYT bread wheats released by the Mexican national research program (Instituto Nacional de Investigaciones Forestales, Agrícolas y Pecuarias— INIFAP) from 1966 to 1988 in the Yaqui Valley. In the second part of the analysis, we estimated the predicted longevity of resistance for varieties in the Yaqui Valley using known data on the year resistance was overcome and a simple actuarial computation. Although the second step was not critical to this particular analysis, it may prove useful in other studies when the longevity of resistance is known for only a sample of varieties or when data permit the development of a fuller econometric model of "duration." In the third step, we conducted a simulation analysis using area distributions and the actual longevity for varieties in the Yaqui Valley, varying parameter values estimated in the first two steps to demonstrate the sensitivity of results to underlying assumptions about their magnitude.[3] In the fourth step, after adding cost data, we computed the internal rate of return on the investment. Each step is described in the sections that follow.

11.3.1. Yield Effects of Specific and Non-Specific Resistance to Leaf Rust

In an experiment conducted and analyzed by Sayre *et al.* (1998), 15 CIMMYT bread wheats released by INIFAP between 1968 and 1988 were grown under farmers' management conditions for four seasons in the Yaqui Valley, with and without fungicide. The regression analysis shown in Table 11.2, which includes control variables related to management and season, information about the known genetic basis of resistance in the varieties, and the age of each variety, shows the estimated effects of the genetic resistance on yield losses from leaf rust. The dependent variable in the regression is the yield difference between protected and unprotected plots, as a percentage of the protected yield.

All 15 bread wheats were highly resistant to leaf rust at the time of release, but all resistance based on specific genes, except that of *Lr19*, has since been defeated. *Lr19* is a specific gene that is still effective in the Yaqui Valley. The *Lr16* complex includes a specific gene (*Lr16*) with at least two other partially effective genes. Alone, *Lr16* confers only moderate resistance; its resistance increases additively in the presence of the other genes (Singh and Huerta-Espino, 1995). The *Lr34* complex includes *Lr34*, which has a partial effect, as well as other non-specific genes. We classified both the *Lr16* and *Lr34* complexes as non-specific resistance for this study.

No more than one of these types of genetic resistance is found in a single variety. The presence in a variety of any one of them—effective specific gene *Lr19*, the *Lr16* complex, or the *Lr34* complex—is associated with an average reduction of about 50 percentage points in yield losses from leaf rust infection under experimental conditions (Table 11.2). These results imply that if either an effective specific gene or a non-specific gene complex is challenged with the same disease pressure in a single season, disease losses are essentially the same. The variable "other non-specific" refers to genes that are not included in the *Lr16* or *Lr34* complexes. The individual effects of

Table 11.2. Effects of Genetic Resistance on Yield Losses from Leaf Rust in 15 Cultivars Released in Mexico, 1968–1988

	Regression coefficient	Standard error
Dependent variable		
Percent of yield lost to leaf rust disease		
Explanatory variables		
Constant	−7.76	
Dummy (1 if trial year 1991, 0 otherwise)	−5.18	2.95
Dummy (1 if trial year 1992, 0 otherwise)	−1.43	1.86
Dummy (1 if trial year 1993, 0 otherwise)	4.47	2.33
Dummy (1 for normal planting date; 0 if late planted)	−15.9*	4.29
Days to anthesis	0.778*	0.207
Days to maturity	0.105	0.196
Age of cultivar in 1996	0.609*	0.122
Presence of *Lr16* complex	−51.0*	3.16
Presence of *Lr19* gene	−50.8*	3.33
Presence of *Lr34* complex	−47.9*	2.39
Number of other non-specific genes	−30.14*	3.86
(Number of other non-specific genes)(squared)	4.73*	1.64
R^2	0.73	

Source: Trial data, Yaqui Valley, described in Sayre *et al.* (1998).
Note: Number of observations=270; * indicates statistically significant at 0.01 level with one-tailed t-test.

these genes are not as great (an average of a 30% reduction in yield losses) and, as suggested by the estimated coefficient of the quadratic term, decrease with each additional gene.

11.3.2. Predicted Longevity of Resistance to Leaf Rust

The regression coefficients reported in Table 11.2 provide an estimate of the difference in percent yield loss from rust in bread wheats with non-specific and specific resistance, once specific genes are no longer effective and under conditions of heavy disease pressure. When combined with information about the known or predicted longevity of specific resistance, the coefficients can be used to estimate the time path of resistance and the benefit streams associated with non-specific resistance.

To predict the longevity of resistance in bread wheats grown in the Yaqui Valley, we developed genetic profiles similar to those used in the analysis of trial data for each of the 38 bread wheat varieties that have been grown by farmers on more than 1% of the area in the Yaqui Valley from 1968 to 1996. These include, but are not limited to, varieties grown in the trials described earlier. Definitions of the effectiveness of resistance were based on pathologists' scoring systems. Resistance was defined as effective as long as the variety maintained useful levels of resistance in most years, showing more infection when disease pressure was heavy but not succumbing. The definition is appropriate for this study because farmers often continue to grow varieties with this level of resistance, even though wheat breeders or pathologists may no longer consider it satisfactory.

To make actuarial predictions of the longevity of rust resistance in varieties, we used non-parametric analysis of frequency distributions for varieties whose resistance has been overcome and those for which the time to failure has not been observed. The expected longevity of rust resistance is calculated as the cumulative proportion of varieties whose resistance has not been overcome by the beginning of the time interval, factoring in uniformly the effects of unobserved duration times. This method provides estimates of longevity that can be compared statistically between types of genetic resistance, and can also be used when longevity of resistance is known only for a sample of varieties grown, rather than for all of them, as in the Yaqui Valley.

These predictions are actuarial in the sense that results are based entirely on observed duration times, and we are unable to use information on related variables to improve predictions. Parametric estimations or a fuller "duration" model analyzing the effects on longevity of resistance of other factors (such as area distributions of varieties, or weather conditions) would be possible only if resistance had been overcome for some of the varieties carrying non-specific resistance.

The analysis demonstrated that by the interval beginning in the third year after release, for all varieties that do not contain non-specific resistance, there is a 70% cumulative probability that a variety's resistance to leaf rust will have failed. None of the varieties carrying the *Lr34* complex have lost their resistance. The expected lifetime of resistance to leaf rust in a variety is only 2.9 years. The null hypothesis that the frequency distributions for longevity of resistance are the same was rejected because none of the varieties with non-specific resistance have lost their resistance. The relationships estimated from the data suggest time paths of yield loss for specific and non-specific resistance as depicted in Figure 11.1.

11.3.3. Summary of Method and Data Used to Simulate Benefits

If actual disease losses over the time period of study were known, the calculation of benefits in terms of yield losses avoided would be straightforward. In year t, m wheat varieties $i = 1,.....m$ are grown by farmers, of which $n \leq m$ carry specific resistance,

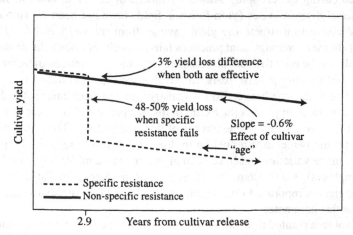

Figure 11.1. Depiction of Yield Time Paths when Resistance to Leaf Rust is Overcome in Bread Wheats Grown in the Yaqui Valley of Mexico from 1970 to 1995.
Note: Percentage points are from Table 11.2.

and the remainder carry non-specific resistance to leaf rust. Variety j, with non-specific resistance, maintains its disease resistant yield of y_{jt} throughout the years that farmers choose to grow it, or its field life. Variety i, with specific resistance, has a disease resistant yield until $t = d_i$, the year when its resistance is no longer effective. This year is specific to each variety. From that year until the end of its field life in year $t = K_i$ (for the period $d_i < t < K_i$), the variety is susceptible to disease loss and its yield is only y_{it}^* $= (1 - \gamma) y_{it}$, where γ is the percentage yield loss for the variety in year t. The output losses from the cultivation of varieties with specific resistance over the time period are the total of annual losses for all susceptible varieties weighted by the areas planted to them. These output losses, when valued, are the "costs" of planting varieties with specific rather than non-specific resistance, or the potential "benefits" of non-specific resistance in terms of yield losses that could have been avoided.

When actual losses are not known, the key parameters in determining the magnitude of expected benefits are: (1) the disease losses for susceptible varieties in any year (γ); (2) the yield level of the varieties with specific resistance (y_{it}); (3) the length of the period from the year when resistance is overcome to the year that farmers no longer choose to grow the variety ($K_i - d_i$); (4) the extent of the area in susceptible varieties (a_{it}^*); (5) the real price of wheat; and (6) the interest rate used to represent the time value of money in the benefits stream. The larger the percentage disease loss for a given yield level, the higher the resistant yield, the more extensive the area grown to the variety after resistance is overcome, or the longer the period it remains in cultivation, the greater the expected cost of specific resistance. The greater the expected cost of specific resistance, the greater the expected benefits of non-specific resistance.

In simulating the benefits stream for non-specific resistance, disease loss parameters and data on the longevity of resistance were combined with area, yield, and price information for bread wheat varieties grown in the Yaqui Valley from 1970 to 1990. The disease loss for any variety in any year (γ) is generally higher in zones of high disease pressure than in others, and much higher in epidemic years when losses cannot be averted during the cropping season by means of chemical control. In the Yaqui Valley, actual disease losses (γ) in farmers' fields have not been measured. The trial data used above to estimate the yield savings from the $Lr16$ and $Lr34$ complexes represents farmers' management practices fairly closely, although the disease pressure in the trials was heavier than the disease pressure now experienced in farmers' fields in the Yaqui Valley during most years. In the absence of actual data on annual disease losses or data on the weather conditions, management practices, and spatial distributions that would allow us to predict the extent of disease pressure from year to year, it is not possible to derive quantitative estimates of the time paths γ_t. The parameter estimate we have for the percentage reduction in disease loss resulting from the non-specific resistance in the varieties that have been grown in the Yaqui Valley (based on $Lr16$ or $Lr34$ complexes) is a constant, γ. We used the estimates from Table 11.2 and other anecdotal figures reported in the literature (see note 1) to establish upper and lower bounds on this parameter.

The total area planted to varieties with specific and non-specific resistance is shown in Figure 11.2. In the Yaqui Valley, since 1990, economic factors unrelated to resistance may have contributed to a rising average age and area-weighted average age of bread wheat varieties grown by farmers. For varieties with specific resistance, this implies a longer period from d_i to K_i. However, bread wheats with specific resistance occupy a generally decreasing percentage of the bread wheat area in the Yaqui Valley. The percentage fluctuates over time, with varietal diffusion paths. The resurgence in area planted to varieties with specific resistance in the mid-1980s is associated with the diffusion of the high-yielding, widely adapted wheat Seri 82 from a cross involving spring and winter wheats.

Since farm-level yield estimates were not available by variety, yields of resistant varieties were calculated algebraically from the average yield data for all bread wheats grown in the Yaqui Valley. Longevity of resistance and field life were known for each of the varieties. The parameter used for area represents the hectares actually planted to varieties with non-specific resistance that would have been planted to susceptible varieties with specific resistance had the research investment in non-specific resistance not been made. We assumed that proportion to be the same as the actual proportion of area in susceptible varieties with specific resistance. This assumption implies that the pattern of area vulnerable to disease loss would be the same for the area planted to varieties with specific and non-specific resistance, even though it is determined by adoption cycles associated with individual varieties. There is little basis for alternative assumptions, however.

As a measure of the worth of the research investment, we used the discounted value of the estimated yield losses that were avoided in the Yaqui Valley from 1970 to 1990 by choosing to invest in non-specific resistance. Discounting the benefits implies that we are viewing the investment decision from the beginning of the time period (1970), as did the decisionmakers in CIMMYT's wheat breeding program when they changed their strategies for incorporating resistance to leaf rust. We used the real rural price of wheat in new pesos (base year = 1994) to account for the effects of inflation on the value of benefits generated in each year. For the sensitivity analysis presented in Table 11.3, we converted the sum of discounted benefits to US dollars at the average 1994 exchange rate.

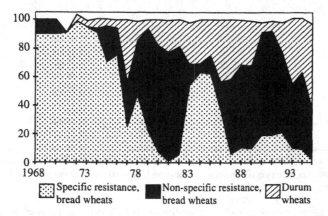

Figure 11.2. Percent of Total Wheat Area by Wheat Type and Type of Leaf Rust Resistance, Yaqui Valley, 1968–1995.

11.3.4. Internal Rate of Return

We have also estimated the internal rate of return on the research investment, or the interest rate that equates the sum of real benefits to the sum of real costs. In this calculation, we assumed a cost roughly equivalent to the salaries and operating expenses of a full-time wheat breeder and a full-time pathologist, or a real cost of US$ 0.5 million per year, beginning in 1968. We conducted several alternative rate of return analyses, with benefit streams beginning after research lags of five years and ten years.

11.4. RESULTS AND SENSITIVITY ANALYSIS

As conceptualized in this paper, the estimated gross benefits of non-specific resistance to leaf rust in bread wheats represent the value of the expected yield losses that were avoided by the decision in 1970 to incorporate non-specific rather than specific resistance into CIMMYT bread wheats. In the base scenario, we used the area distributions and yields as calculated above, with a very conservative estimate of a 9% yield loss and an

Table 11.3. Estimated Gross Benefits of Non-Specific versus Specific Resistance to Leaf Rust in Wheats Grown in the Yaqui Valley of Mexico, 1970–1990

Parameter[a]		Benefits (million 1994 US$)	Average annual % of total value of wheat production[b]	Change in benefits from base (US$)	Percentage change in benefits from base
Base					
γ		9	16.90	0.547	
r		10			
a^*		35			
d_i	Varies, mean=3				
K_i	Varies, mean=5				
Sensitivity					
γ		19	35.60	1.16	19	111
		39	73.10	2.37	56	333
		49	91.90	2.98	75	444
r		1	39.50	1.28	23	134
		15	11.70	0.38	–$5	–30.9
a^*		36	26.60	0.86	10	58
		40	62.40	2.02	46	270
d_i		4	14.60	0.47	–2	–13.7
K_i	Mean=5[c]		24.10	0.78	7	42.8

a γ is the average percent yield loss due to leaf rust, r is the interest rate, a^* is the percentage of bread wheat area in cultivars with specific resistance that is no longer effective, averaged over 1970–1990; d_i is the number of years from release until resistance is no longer effective; and K_i is the field life of the cultivar. Both d_i and K_i vary by cultivar.

b Average percentage of total value of wheat production in 1994 US dollars at that year's exchange rate, planted area, and with wheat yields of 5 t/ha.

c Increased by one year for each cultivar, at the percentage of area recorded in final year.

interest rate of 10%. Under these conservative assumptions, the worth in 1970 of the benefits generated in the Yaqui Valley alone over 1970–1990 was, expressed in 1994 real terms, US\$ 17 million. As a point of reference, this amount represents, when annualized over the 1970–1990 period, slightly over 0.5% per year of the 1994 value of wheat production in the Yaqui Valley (Table 11.3).

The figures in Table 11.3 report the benefits for the base scenario and how the benefits change with changes in assumptions about parameter values. They can be interpreted as the incremental effects of changes in one parameter value on benefits, holding other parameter values constant.

Estimated benefits in the baseline scenario were calculated with the actual longevity of resistance and lifetimes of bread wheats in farmers' fields in the Yaqui Valley. The area in bread wheats with specific resistance that had broken down averaged 35% of the total bread wheat area (a^*) over 1970–1990. The area planted to susceptible varieties is likely to be considerably lower in the Yaqui Valley than in many other regions of the developing world. An increase of only 1% in the area planted to susceptible varieties increases the benefits by 58%.

Over the time period, the area in susceptible varieties is a function of the time until their resistance is overcome (d_i) and their lifetime in cultivation (K_i). Although heavy disease pressure in the Yaqui Valley means that resistance is overcome quickly, farmers also change their bread wheat varieties rapidly, and the average lifetime of a variety in cultivation is only five years. Increasing the longevity of resistance by only one more year among the Yaqui Valley varieties decreases the benefits by about 14%, because it reduces the average area in susceptible varieties over the time period. When the lifetime in cultivation for susceptible bread wheats is expanded by only one year at the percentage area recorded for their final year in actual cultivation, benefits are 43% higher. The key time period for the size of benefits is the span ($K_i - d_i$), which was only two years on average in the Yaqui Valley during the study period. This period of time is undoubtedly brief for the Yaqui Valley compared to other wheat-producing areas in developing countries.

The disease loss parameter (γ) used in the baseline scenario was reduced from that estimated in the analysis of trial data, to represent lower disease pressure in farmers' fields in most years. During the time period under study, yield losses of over 40% occurred only in the epidemic year of 1977 (Dubin and Torres, 1981), although they occurred more frequently and with great severity in the 1930s and 1940s for stem rust (Borlaug, 1968). In the 1977 epidemic, yield losses were mitigated by a massive, well-organized campaign of chemical control. Calculations of benefits are very sensitive to assumptions about average disease losses, and with the highest disease losses, benefits are US\$ 92 million (in real 1994 terms). The highest disease loss scenario would represent an unlikely situation (chronic, severe epidemics).

Benefits are also particularly sensitive to the level of the interest rate r, or assumptions about how we value money over time. Which rate is appropriate is the subject of extensive debate in both the applied and theoretical economics literature. The debate centers on which concept of the value of capital to use. For example, if r is the "opportunity cost of capital" in an economy, it represents the return on the marginal investment that uses the last of the available capital. This r is meant to reflect "the choice made by the society as a whole between present and future returns, and, hence, the amount of total income the society is willing to save"(Gittinger, 1982: 314). For developing countries, a range of 8–15% in real terms is often assumed. Applying a 15% rate in this example causes a fairly large reduction in estimated gross benefits.

Another possible r is the "social time preference rate," which reflects the idea that society has a longer time horizon than individuals and implies that a lower r be used for public projects than for private projects. In this example, an r of 1%, which implies a higher discount factor, generates a 134% increase in estimated gross benefits.

While the base scenario appears to generate fairly modest gross benefits, these benefits are sufficient to generate fairly sizable rates of return on capital investment. Even with heavy cost penalties in the earlier years, a lag of ten years in research benefits, and the most conservative disease loss assumption, the rate of return on capital is about 13%—well within the range recommended for use as an acceptable "opportunity cost of capital" for projects under consideration by the World Bank. The rate of return used by the World Bank is arguably high for investments made by a public research institution like CIMMYT. A research lag of five years and the most conservative disease loss assumptions generates an internal rate of return of nearly 40% (Figure 11.3).

11.5. CONCLUSIONS

The results present a fairly strong case that the economic benefits of CIMMYT's strategy to incorporate non-specific rather than specific resistance to leaf rust in bread wheat have been substantial. While the numerical values of the estimated benefits for the Yaqui Valley are sensitive to assumptions about underlying parameter values, they remain large enough that when very conservative cost and benefit assumptions are used, the internal rate of return to capital invested satisfies stringent investment criteria. The conditions that determine a favorable rate of return on the investment are likely to hold in many favorable environments for growing wheat in developing countries, where

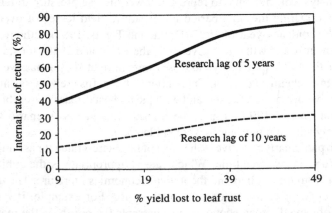

Figure 11.3. Internal Rate of Return to Research Investment in Non-Specific vs. Specific Resistance to Leaf Rust in Modern Bread Wheats, Yaqui Valley of Mexico.

farmers change varieties slowly because of delays in varietal release or other economic factors related to the rate of adoption, disease pressure is heavy, and the costs of treating outbreaks is high.

How do these results compare to what might be estimated as a global rate of return to investment for CIMMYT's mandate area? Several factors contribute to an underestimation of benefits in this study. First, the research investment in non-specific resistance to leaf rust applies to an estimated area of 45–50 million hectares of wheat in the developing world, as compared to the 150,000 ha in the Yaqui Valley. The total area in susceptible varieties is likely to be higher in other countries of the developing world than in the Yaqui Valley. The Yaqui Valley is the location of one of CIMMYT's major experiment stations and the fields of its farmers are some of the first testing grounds of CIMMYT wheats outside experiment stations. Although disease pressures are greater there than in many other parts of the developing world, the speed with which farmers replace varieties curtails the time period over which the benefits accrue for each variety. For some production environments, however, longer research lags, lower disease pressure, and lower rates of adoption of varieties with non-specific resistance would reduce the magnitude of the benefits per unit of area below those calculated for the Yaqui Valley.

In enlarging the scale of the analysis from the Yaqui Valley to CIMMYT's mandate area, costs are not likely to increase proportionally to benefits. Although the costs of the international testing network and costs borne by national programs have been excluded in this analysis, the cost streams used here represent heavier investments in early years than was actually the case, and they include the full cost at CIMMYT headquarters of the rust research program. If incremental costs associated with breeding for non-specific rather than specific resistance were calculated, the costs of breeding for specific resistance would likely be greater than those associated with non-specific resistance.

Our calculations have omitted a dimension of benefits from non-specific resistance that is difficult to measure but may be important to recognize. Output losses from rust include both incremental, annual losses and the major losses incurred by epidemics. How socially important these losses are depends not only on their absolute magnitude, but the role of wheat production in the national economy, the attitude of that society toward risk, the time horizon, and other considerations influencing the valuation of the yield loss. For some farmers and societies, the real costs of these losses—and especially epidemics—can be great because of the extent to which they rely on the wheat crop. Large crop losses may imply price increases which are passed to consumers, or unforeseen imports purchased at world market prices which may not be favorable. Some farmers and some societies therefore place a "premium"—an added value—on avoiding disasters. Modern epidemics may require treatment with fungicide and large-scale, well-coordinated mobilization campaigns. Heavy chemical treatments may have secondary health costs. Non-specific resistance effectively changes the yield distribution by reducing the probability of falling into the lower tail. By reducing the probability of "disaster," it benefits farmers and societies for which such outcomes are "disastrous."

Acknowledgments

Ken Sayre, agronomist with the CIMMYT Wheat Program, provided the experimental data for this analysis. The authors gratefully acknowledge the insights of Jesse Dubin, Paul Heisey, Prabhu Pingali, Sanjaya Rajaram, and Greg Traxler. Efrén del Toro, who was with the Wheat Program when this research was conducted, constructed Table 11.1. Pedro Aquino and Dagoberto Flores, CIMMYT Economics Program, shared their knowledge of wheat production in the Yaqui Valley. A discussion of these results, oriented towards plant pathologists rather than economists, is forthcoming in *Plant Disease.*

Notes

1   There are many historical accounts of the economic importance of wheat rusts, although in most cases the relationship to prices, wheat output levels, and wheat imports is reported anecdotally. Estimated production losses in some of these accounts range from 10% to 50% of the crop, depending on the year and the size of the area considered. These include: for the Asian subcontinent, Barclay (1892), Howard and Howard (1909), Nagarajan and Joshi (1975, 1985), M.A. Khan (1987), Nagy (1984); for other regions, including Europe and North America, Roelfs and Bushnell (1986), Stakman and Harrar (1957); for Mexico, Borlaug (1968), Dubin and Torres (1981). See also Oerke *et al.* (1994). The cereal rusts are a fungal disease with "worldwide" occurrence and are characterized by "frequent severe epidemics" and "huge annual losses" (Agrios, 1988: 20).

2   The novelty of using this approach in the 1970s is illustrated by the reference to non-specificity in the famous study on genetic vulnerability that was published by the National Research Council in 1972, which states that "there is evidently another biological mechanism that limits the potential impact of these diseases [of wheat] on yield losses . . . . Everyone talks about it but few try to do it . . . . Non-specificity is poorly understood and not used in an orderly manner in breeding programs primarily because we do not know how to select for it, we have no knowledge of the genetic mechanisms that control it, and worse, we do not know how effective it would be in reducing losses caused by many of the diseases of wheat"(p. 141).

3   Originally, with the same genetic information, we also attempted to estimate the effect of carrying non-specific resistance genes on wheat yields through estimation of a production function with regression analysis of farm survey data (Flores, 1997). As may be expected, with this procedure we were unable to capture the effects of specific genes on yield because of overwhelming effects of management and environmental variables. We then decided to conduct a simulation analysis.

References

Agrios, G. N. 1988. *Plant Pathology.* Third edition. New York: Academic Press.

Barclay, A. 1892. Rust and mildew in India. *The Journal of Botany* 30: 1–8.

Borlaug, N. E. 1968. Wheat breeding and its impact on world food supply. Paper presented at the third International Wheat Genetics Symposium, 5–9 August, Canberra, Australia.

Brennan, J. P., and D. Byerlee. 1991. The rate of crop varietal replacement on farms: Measures and empirical results for wheat. *Plant Varieties and Seeds* 4: 99–106.

Brennan, J. P., G. M. Murray, and B. J. Ballantyne. 1994. *Assessing the Economic Importance of Disease Resistance in Wheat.* NSW Agriculture, Agricultural Research Institution, Wagga Wagga. Final Report to the Grains Research and Development Corporation, Australia.

Byerlee, D., and P. Moya. 1993. *Impacts of International Wheat Breeding Research in the Developing World, 1966–90.* Mexico, D.F.: International Maize and Wheat Improvement Center (CIMMYT).

Byerlee, D., and G. Traxler. 1995. National and international wheat improvement research in the post-green revolution period: Evolution and impacts. *American Journal of Agricultural Economics* 77 (2): 268–278.

Caldwell, R. M. 1968. Breeding for general and/or specific plant disease resistance. In K. W. Finlay and K. W. Shepherd (eds.), *Proceedings of Third International Wheat Genetics Symposium.* Canberra, Australia: Australian Academy of Sciences.

Collins, M. 1995. The economics of productivity maintenance research: A case study of wheat leaf rust resistance breeding in Pakistan. Ph.D. thesis, University of Minnesota, St. Paul, Minnesota.

Dubin, H. J., and E. Torres. 1981. Causes and consequences of the 1976–77 wheat leaf rust epidemic in northwest Mexico. *Annual Review of Phytopathology* 19: 41–49.

Evenson, R. E. 1998. Crop-loss data and trait value estimates for rice in Indonesia. In R. E. Evenson, D. Gollin, and V. Santaniello (eds.), *Agricultural Values of Plant Genetic Resources*. Wallingford: CAB International.

Flores, D. V. 1997. *Analysis of Yaqui Valley Wheat Production—On-Farm Diagnostic Study, 1981–1996*. Economics Program internal document. Mexico, D.F.: International Maize and Wheat Improvement Center (CIMMYT).

Gittinger, J. P. 1982. *Economic Analysis of Agricultural Projects*. Second Edition. Published for the Economic Development Institute of the World Bank. Baltimore: Johns Hopkins.

Heim, M. N., and L. Blakeslee. 1986. *Biological Adaptation and Research Impacts on Wheat Yields in Washington*. Pullman, Washington: College of Agriculture and Home Economics, Washington State University.

Heisey, P. W., and J. P. Brennan. 1991. An analytical model of farmers' demand for replacement seed. *American Journal of Agricultural Economics* 73: 1044–1052.

Howard, A., and G. L. C. Howard. 1909. *Wheat in India: Its Production, Varieties and Improvement*. Calcutta: Thacker, Spink and Company, for the Imperial Department of Agriculture in India.

Khan, M. A. 1987. *Wheat Variety Development and Longevity of Rust Resistance*. Lahore: Government of the Punjab Agriculture Department.

Kilpatrick, R. A. 1975. *New Wheat Varieties and Longevity of Rust Resistance, 1971–5*. ARS-NE-4. Beltsville: Agricultural Research Service, US Department of Agriculture.

Lupton, F. G. H. 1987. History of wheat breeding. In F. G. H. Lupton (ed.), *Wheat Breeding: Its Scientific Basis*. London: Chapman and Hall.

Macindoe, S. L., and C. W. Brown. 1968. *Wheat Breeding and Varieties in Australia*. Science Bulletin No. 76. Third edition. Sydney: New South Wales Department of Agriculture.

Morris, M. L., H. J. Dubin, and T. Pokhrel. Returns to wheat breeding research in Nepal. 1994. *Agricultural Economics* 10: 269–282.

Nagarajan, S., and L. M. Joshi. 1975. An historical account of wheat rust epidemics in India and their significance. In J. G. Manners (ed.) *Cereal Rusts Bulletin* 3(2): 29–33.

Nagarajan, S., and L. M. Joshi. 1985. Epidemiology in the Indian Subcontinent. In A. P. Roelfs and W. R. Bushnell (eds.), *Diseases, Distribution, Epidemiology, and Control*. Vol. 2 of *The Cereal Rusts*. London: Academic Press.

Nagy, J. G. 1984. The Pakistan agricultural development model: An economic evaluation of agricultural research and expenditures. Ph.D. thesis, University of Minnesota, St. Paul, Minnesota.

National Research Council, Committee on Genetic Vulnerability of Major Crops. 1972. *Genetic Vulnerability of Major Crops*. Washington, DC: National Academy of Sciences.

Oerke, E.-C., H.-W. Dehne, F. Schönbeck, and A. Weber. 1994. *Crop Production and Crop Protection: Estimated Losses in Major Food and Cash Crops*. Amsterdam: Elsevier.

Plucknett, D. L., and N. J. H. Smith. 1986. Sustaining agricultural yields. *BioScience* 36(1): 40–45.

Priestley, R. H., and R. A. Bayles. 1988. Contribution and value of resistant varieties and disease control in cereals. In B. B. Clifford and E. Lester (eds.), *Costs and Benefits of Disease Control*.

Rajaram, S., R. P. Singh, and E. Torres. 1988. Current CIMMYT approaches to breeding for rust resistance. In N. W. Simmonds and S. Rajaram (eds.), *Breeding Strategies to the Rusts of Wheat*. Mexico, D.F.: International Maize and Wheat Improvement Center (CIMMYT).

Rajaram, S., R. P. Singh, and M. van Ginkel. 1996. Approaches to breed wheat for wide adaptation, rust resistance and drought. In R. A. Richards, C. W. Wrigley, H. M. Rawon, C. J. Rebetzke, J. L. Davidson and R. I. S. Brettell (eds.), *Proceedings of the 8th Assembly of the Wheat Breeding Society of Australia*, 29 Sept.–4 Oct. 1996. Canberra, Australia.

Rejesus, R., M. Smale, and M. van Ginkel. 1997. Wheat breeders' perspectives on genetic diversity and germplasm use: Findings from an international survey. *Plant Varieties and Seeds* 9: 129–147.

Roelfs, A. P. 1988. Resistance to leaf and stem rusts in wheat. In N. W. Simmonds and S. Rajaram (eds.), *Breeding Strategies for Resistance to the Rusts of Wheat*. Mexico, D.F.: International Maize and Wheat Improvement Center (CIMMYT).

Roelfs, A. P., and W. R. Bushnell (eds.). 1986. *Diseases, Distribution, Epidemiology, and Control*. Vol. 2 of *The Cereal Rusts*. London: Academic Press.

Sayre, K. D., R. P. Singh, J. Huerta-Espino, and S. Rajaram. 1998. Genetic progress in reducing losses to leaf rust in CIMMYT-derived Mexican spring wheat cultivars. *Crop Science* 38: 654–659.

Singh, R. P., and H. Jesse Dubin. 1997. Sustainable control of wheat diseases in Mexico. In *Memorias de 1er Simposio Internacional de Trigo, 7–9 April, 1997, Cd. Obregón, Sonora, Mexico*. Mexico, D.F.: International Maize and Wheat Improvement Center (CIMMYT).

Singh, R. P. and J. Huerta-Espino. 1995. Inheritance of seedling and adult plant resistance to leaf rust in wheat cultivars Ciano 79 and Papago 96. *Plant Disease* 79(1): 35–38.

Smale, M. 1996. *Understanding Global Trends in the Use of Wheat Diversity and International Flows of Wheat Genetic Resources*. Economics Working Paper 96–02. Mexico, D.F.: International Maize and Wheat Improvement Center (CIMMYT).

Stakman, E. C., and J. G. Harrar. 1957. *Principles of Plant Pathology*. New York: The Ronald Press Company.

Vanderplank, J. E. 1963. *Plant Diseases: Epidemics and Control*. New York: Academic Press.

V

Policies and Genetic Resource Utilization

12 INSTITUTIONAL CHANGE AND BIOTECHNOLOGY IN AGRICULTURE: IMPLICATIONS FOR DEVELOPING COUNTRIES

D. Zilberman, C. Yarkin, and A. Heiman

12.1. INTRODUCTION

Agriculture in the modern era has been characterized by high rates of technological change (Cochrane, 1993; Schultz, 1964). Over the past century, successive waves of innovation have dramatically altered agricultural production systems. These innovations include several types of technologies: mechanical (tractors, combines, harvesters); chemical (fertilizers and synthetic pesticides); irrigation-related (tubewells, tiledrains, sprinkler and drip irrigation); and biological (improved varieties and hybrids developed through selective breeding). While each of these technological changes has primarily affected one particular feature of the production system, all of them have generally entailed significant modifications in other components of the system. The green revolution, for example, introduced new varieties and hybrids but required more intensive use of irrigation, fertilizer, and pest control.

Recently we have witnessed a new wave of innovations based on biotechnology. This chapter assesses some of the important institutional features that affect the way agricultural biotechnology is introduced and adopted. We analyze key issues associated with the introduction of biotechnology in developing countries by examining the lessons learned in developed nations, particularly in the United States. We define "biotechnology" to encompass the wide array of innovations that stem from modern molecular biology, including recombinant DNA techniques (genetic engineering).

The next section of this chapter identifies the major categories of biotechnology innovations that affect agriculture, including innovations that enhance the supply of products, improve the quality of agricultural commodities, and expand the range of products. Section 12.3 describes the nature of the private- and public-sector institutions that share responsibility for researching, developing, producing, and marketing the products of biotechnology. We explore alternative mechanisms to finance biotechnology research among public-sector institutions, the transfer of technology from public- to private-sector institutions, evaluation and control of intellectual property rights associated with biotechnology, and major issues related to the marketing and sale of biotechnology products to final consumers.

Section 12.4 focuses on how the issues discussed in the previous sections may affect developing countries, with implications for the role of the international agricultural research centers. We propose alliances among researchers and producers in developing countries with organizations in developed countries so that biotechnology can be introduced and used in the developing world. We also refer to issues of intellectual property rights, pricing, and access to biotechnology products in developing countries.

12.2. CATEGORIES OF AGRICULTURAL BIOTECHNOLOGY

Biotechnology has the potential to produce a vast number of products. In agriculture, biotechnology can be grouped into technologies that increase the supply of products, those that enhance product quality, and those that create new products.

Technologies that increase the supply of products include innovations that enhance the efficiency of input use. Biotechnology provides tools to accelerate the speed and productivity of traditional crop and livestock breeding systems through more efficient use of inputs such as feed, fertilizer, and water. Bovine growth hormone, one of the first major biotechnological innovations introduced in agriculture, is one example. Another example is crop varieties that are modified using gene-splicing techniques. The adoption of such varieties occurs first and most often in locations where the efficiency-enhancing effect is most substantial. Khanna and Zilberman (1997) found that factors inducing adoption include a rise in output price, an increase in input prices when the variety reduces input use, and environmental regulations aimed at curbing residues or externalities, such as a tax on pollution. The introduction of varieties that enhance the efficiency of input use expands supply, both by increasing output at present production locations (the intensive margin) and expanding the area where a crop or livestock can be produced (the extensive margin). Of course, if technology has a discernible supply-shifting effect and the product has a relatively low demand elasticity, adoption of a new variety may result in a reduced output price. This will lead to a secondary set of adjustments until an equilibrium is reached.

Biotechnology can also affect the supply of agricultural products by reducing crop damage from pests and diseases. The most widespread application of biotechnology in agriculture has been in pest control. Biotechnology techniques may provide solutions to disease and pest problems not addressed by existing methods of pest control, increase the reduction in pest damage in comparison to current pest control methods, substitute for pest control, or in some cases complement current practices by enhancing their effectiveness. For example, in 1997 cotton and soybeans modified to express the bacterial toxin *Bt* (*Bacillus thuringiensis*) were used on millions of acres in the United States to reduce disease and insect damage. When biotechnology-based pest control substitutes for an existing pesticide, its adoption, use, and value to the manufacturer may be enhanced significantly by environmental regulations restricting the use of the existing treatment.

Other biotechnology innovations improve product quality. Over the years, the relative prices of agricultural commodities have declined dramatically, and the source of "agricultural problems" in most developed and in some developing countries has been an excess of agricultural commodities. Farmers and agribusinesses have realized that as consumers' incomes increase, they are willing to pay higher prices to purchase higher quality products. An important strategy to improve farm income is to increase the value added and improve the quality of agricultural products. Genetic manipulation may be extremely helpful in developing new varieties with higher value added by enhancing desirable features of food products, extending shelf life, and suppressing undesirable characteristics.

Studies using hedonic techniques (Parker, Zilberman, and Castillo, 1998) demonstrate that improving the quality of products (e.g., sugar content or color of certain fruits and vegetables) and increasing their availability throughout the year may raise prices of these products substantially in developed countries. Genetic manipulation can be effective in improving taste and multiseasonal availability of agricultural products. Gains may also be obtained by increasing the nutritional content of certain food items or reducing elements that are detrimental to health (reducing the fat content in meat and cholesterol in eggs are two examples of activities that may generate added value). Value may also be obtained from fortifying food products with micronutrients, minerals, and other substances with positive or even therapeutic health effects. Genetic manipulation may improve the physical structure of certain agricultural commodities to make them easier to harvest and handle.

Although biotechnology can be used to improve the quality of food, some consumers may be apprehensive about its use. The negative connotations associated with biotechnology processes could actually diminish the added value of the characteristics they enhance. Negative perceptions can be reduced by educating the public about how biotechnology fits into the overall range of agricultural technologies and by ensuring that appropriate health and safety regulations are in place.

Finally, biotechnology can create new products, expanding the range of products available from agricultural systems. Crop and livestock systems can be modified to yield fine chemicals (e.g., valuable nutrients, chemical feed stocks, polymers), oils for human and industrial uses, medical and pharmaceutical products, and new sources of energy. Biotechnology will expand our current uses of domesticated crops and livestock and increase the number of organisms used in agricultural activities. We already have farms that raise algae, a number of fish species, and forest products, and the range of agricultural products will increase even more. Agriculture in the next century will be more than a source of food and fiber: it will become a major supplier to the pharmaceutical, chemical, and consumer goods industries.

12.3. STAGES IN PRODUCT DEVELOPMENT AND ASSOCIATED INSTITUTIONAL ARRANGEMENTS

Figure 12.1 presents a cycle of activities and new institutional relationships associated with agricultural biotechnology. These activities can be divided into two major stages. The first stage consists of activities that generate and produce agricultural inputs (e.g., seed, genetic material) and lead to the establishment and enforcement of intellectual property rights. These activities include motivation, research, development, registration, and the marketing and adoption of agricultural inputs. The second stage consists of activities associated with agricultural output, such as production, marketing, and consumption. There are important feedback relationships within the system. For example, consumer preferences and environmental regulations may motivate research that will lead to new agricultural biotechnology.

12.3.1. Universities and Technology Transfer

In the past, most major innovations were originated by practitioners. Academic faculty discovered basic principles of chemistry, physics, biology, and medicine. These principles provided the base from which applied knowledge and new technologies emerged, but the actual application was generally left to others. Around the turn of the century, scientific entrepreneurs such as Thomas Edison and Alexander Graham Bell created some of their most significant inventions in private laboratories, and the patents they obtained provided the foundation for new industries. Research in the laboratories of large companies, including DuPont, AT&T, IBM, Ford, and Kodak, has inspired many of the commercial innovations introduced during the 20th century. The United States government established research laboratories to pursue research and development (R&D) for the military and agriculture. Over the past 50 years, academic researchers have contributed greatly to the evolution of new technologies and industries. This is also true in the case of biotechnology.

An institutional structure that has facilitated the development and utilization of commercially viable innovations discovered by university and government scientists is the office of technology transfer. The first such office was established at the Massachusetts Institute of Technology in 1944. In recent years, most major research universities and institutes have established their own technology transfer offices. Postlewait, Parker, and Zilberman (1993) and Parker, Zilberman, and Castillo (1998) have studied the economics and performance of technology transfer in the United States, leading to five general conclusions.

First, the central objective of an office of technology transfer is to increase the utilization of innovations developed by universities or institutes. A private company will not develop products invented by universities unless it has established property rights to the patents. The office of technology transfer works with scientists to identify innovations with potential for patents; searches for clients who will pay to file for patent protection (and sign an exclusive license) or who will license non-exclusive rights to existing patents; and arranges for the signing, monitoring, and enforcement of technology transfer agreements. To minimize the likelihood that a company will strategically buy patent rights to suppress innovations that may compete with its products, technology transfer agreements stipulate that licensees should demonstrate some effort to use the rights to develop a product. In most cases, licensees may place only minimal limits (60–90 days) on publishing research results relating to the discovery.

A second feature of the economics and performance of technology transfer is that a handful of innovations may garner most of the revenues from technology transfer. The top ten money-making patents in 1996 generated about 70% of the revenues created by university technology transfer. Alston and Pardey (1996) have also shown that most of the benefits of agricultural research and development are captured by a relatively small number of innovations. Proceeds from technology transfer agreements are shared among the universities, the professors involved, and occasionally their departments.

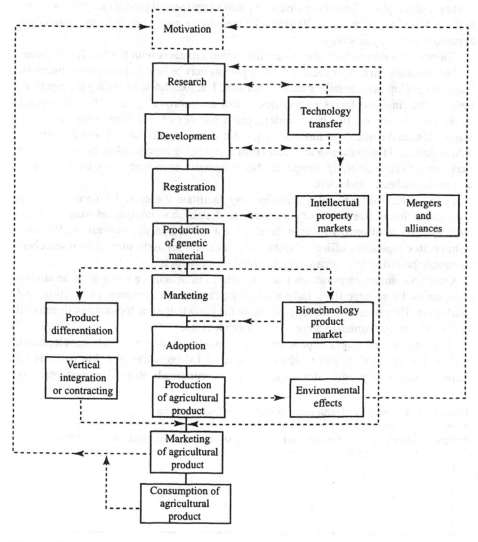

Figure 12.1. Activities Associated with Agricultural Biotechnology.

This three-way sharing is necessary because not all faculty discover commercially valuable inventions, and only a small subset of technology transfer agreements generate large revenue streams. The procedure of allocating proceeds to departments rather than individuals addresses the issue that some technology transfer agreements have made a few researchers very rich. It also recognizes that research teams have benefited from departmental resources.

Third, proceeds from technology transfer agreements may cover only a small fraction of the cost of public research. Earnings from technology transfer are less than 1% of the annual budget for most universities; even in the most successful universities, they are less than 3% of the university budget. In most cases, the royalty rate is between 1% and 5% of income generated annually by an innovation and is of brief duration. On average, developing a laboratory discovery into a commercial product requires 8 years, and the life of a patent life is 20 years. Universities earn revenues from a patent for approximately 12 years only.

Fourth, the value of the rights to an innovation increases with further development of the innovation. In many cases, university patents may be licensed at a pre-commercial stage of development, entailing substantial risk. The likelihood of selling the rights, as well as the financial terms of payments, can be improved greatly by subsequent development. Universities have undertaken a number of activities meant to refine innovations and make them more attractive for commercialization. These include the establishment of foundations and venture capital funds, business incubators, and research parks on or near university campuses. No systematic economic assessment of these activities has been conducted.

Fifth, offices of technology transfer may facilitate the establishment of startup companies that utilize university innovations. Established companies may prefer to avoid the risk or effort involved in developing a new university innovation. Working with venture capitalists, offices of technology transfer may assist university researchers in establishing startup companies to develop new products.

Over time, major corporations may buy shares in or acquire some of these startup companies. These large firms have a relative advantage in licensing, production, and marketing. They license or directly purchase the knowledge and technology developed by the startup companies to augment their product line.

There are several stages of product development, including research, development, registration, production, and marketing. Table 12.1 depicts five possible patterns for different stages of product development and commercialization. The patterns vary

Table 12.1. Alternative Mechanisms for Product Introduction

Pattern	Research	Development	Registration	Production	Marketing
1	U	I	M	M	M
2	U	M	M	M	M
3	U	I	I	M	M
4	M	M	M	M	M
5	I	I	M	M	M

Note: U = universities, I = startups, and M = multinationals.

depending on the kind of organization that undertakes each stage of product introduction: universities (U), startups (I), and major corporations (M). One possible avenue for product introduction is that universities implement research, startups conduct the development and registration, and major corporations execute marketing and production. Another option is to have universities perform research; startups develop the product; and major corporations register, market, and produce the product.

In one of the most prevalent patterns, a major corporation assumes sole responsibility for all stages of product development. This pattern dominates when there is little or no funding for public research, and it leads to fewer innovations and slower technological progress. Corporations may be uninterested in researching and developing product lines that compete with products for which they have monopoly power, and they are more risk-averse than governments and society at large. Furthermore, they take into account only the producer surplus of new technologies and not the consumer surplus. Corporations underinvest in innovations relative to the social optimum.

For these reasons, complete reliance on the private sector for R&D is likely to result in underinvestment. Public investment in academic research produces discoveries that can be transferred to the private sector via offices of technology transfer. An office of technology transfer may work with venture capitalists to establish startups to undertake the development of new products. Public investments can thus increase competitiveness in the marketplace and enhance the rate of innovation. Further research is needed to determine the optimal allocation of resources to public-sector research and to quantitatively evaluate the patterns of collaboration between venture capitalists and the public sector.

Technology transfer agreements provide one avenue for collaboration between private- and public-sector institutions in developing new technologies, although there are several other important formal and informal mechanisms of collaboration, including contracts and grants, gifts, consulting relationships, educational contracts, and equipment sharing and rental arrangements. Wright (1985) has provided an economic foundation for determining the optimality of various arrangements. Clearly, companies expand their human capital and sometimes their physical capital by purchasing or renting university talent or equipment. Similarly, universities finance some of their growth by obtaining resources from collaborating private enterprises. The extent to which this arrangement distorts the university's agenda and research objectivity may be questioned, but significant benefits accrue to both parties.

12.3.2. Intellectual Property Rights, Product Design, and Pricing
Biotechnology discoveries can be divided into process discoveries (for example, a new technique of genetic manipulation) and product discoveries (such as a gene sequence that generates resistance against a specific pest). The patent system and plant protection legislation have been expanded to enable discoverers of new genetic procedures and relationships to own the property rights of their discoveries.

The use of intellectual property rights and patent law in biotechnology must be refined, however, to promote efficient outcomes and discourage abuse. For example, it is not clear how broad the intellectual property rights coverage should be for product discoveries. If a researcher discovers that a particular gene sequence protects certain crop varieties against a disease, should he/she have the intellectual property right to use this sequence for disease protection in all varieties of the same crop, as well as in other plants? To what extent should he/she prove the performance of this sequence in other varieties/species in order to obtain broad intellectual property rights coverage? If intellectual property protection is too limited, it may significantly reduce the incentives for discovery. But if protection is too broad, it may provoke excessive concentration of market power.

Related research issues include the extent to which exclusive rights for new discoveries should be given and the price of these rights. We may find that there is a difference between social and private optima. The social optimum would be defined by the set of prices and royalties that maximize the overall economic benefits to society, including those enjoyed by consumers and producers. The private optimum is given by the prices and royalties that maximize the benefits of individuals or organizations who own the patents or the rights to use the patents.

Biotechnology and the intellectual property arrangements it involves will dramatically affect the design and price of genetic materials. Producers of genetic materials (seed companies) will have a menu of product characteristics and processes that they can choose from in designing new seeds. They will be able to design products with very detailed specifications, either to be sold in seed markets or supplied at the request of agricultural producers. Seed prices will reflect the cost of the intellectual property rights over the discoveries used to produce the seed. Owners of exclusive rights to a valuable product characteristic may have significant monopoly power, and most of the economic surplus associated with this characteristic will be extracted. Extremely high prices or inability to obtain appropriate genetic materials may heighten economic disparities among farmers.

12.3.3. Vertical Integration and Contractual Arrangements in Markets for Differentiated Products

Agricultural output consists of two broad classes of goods: commodities and differentiated products. When agricultural goods are differentiated products, there is a much greater tendency for contractual arrangements to be established among producers, input suppliers, and buyers of the product. Often production is vertically integrated—the functions of input supply, output production, and marketing are all orchestrated by the same organization.

Biotechnology is likely to intensify and accelerate product differentiation in agriculture when innovations result in new, higher quality agricultural products. Many of these differentiated products will be produced under contractual arrangements or vertical integration of operations. The pressure to integrate that results from quality differentiation is more likely to occur in the production of fruits, vegetables, flowers, fine chemicals, and oils than in agricultural commodities such as rice, wheat, and maize.

The coordination needed to introduce differentiated products into a market is a major reason for vertical integration and contractual arrangements in producing these products (Williamson, 1985). Shelf space must be secured, and transport and storage must be negotiated. Increased uncertainty and transaction costs are likely to reduce the attractiveness of a new technology to farmers. Agribusinesses that invest in creating the genetic materials necessary to produce differentiated products will generally want to ensure that they have growers by signing contracts, and they will want to guarantee that they have markets for their final products by signing agreements with food retailers.

Several patterns have emerged in the development of differentiated product markets resulting from innovations in agricultural biotechnology. In the first, multinationals obtain rights to genetic materials of differentiated biotechnology products, contract with farmers for production, and market the products themselves. Suppose the research that provides the basis for differentiated products is conducted by a public research institution. Agribusiness firms may acquire the rights to develop these discoveries either directly through technology transfer or indirectly through contracting with or purchasing the startup companies holding the property rights of interest. These companies may produce genetically improved seeds, contracting with farmers to grow them and marketing the final products. An example of this arrangement is the genetically engineered Flavr Savr™ tomato developed by Calgene. Flavr Savr™ seed is produced through contracts with growers and sold by companies associated with Monsanto.

A second pattern is for corporations specializing in consumer goods to sell differentiated products of agricultural biotechnology. Companies specializing in food production and marketing may either develop subsidiaries with the capacity to produce genetically engineered products or contract with agricultural suppliers who produce the desired inputs. Many of these companies are already engaged in contract farming and enjoy a competitive edge because of their experience in marketing differentiated products (examples may be drawn from Proctor and Gamble, Campbell Soups, and Frito-Lay).

A third pattern occurs when farm cooperatives or commodity groups purchase the right to develop differentiated agricultural products. Cooperatives or farmer groups may contract with companies specializing in biotechnology research, development, and production to produce a seed product and then produce and sell the products themselves. Alternatively, agribusiness firms owning the rights to innovations of differentiated products may produce a seed and then contract with a cooperative or a group of farmers to produce a product that the firms will sell.

In a fourth pattern, large agribusinesses or cooperatives support public institutions through contracts, grants, or gifts to engage in research resulting in discoveries that can be developed into differentiated products. California flower growers, for example, have paid university researchers to conduct studies to improve a flower's color, scent, and durability. These sorts of arrangements are more likely to occur as the uncertainty about research capability declines.

12.3.4. Alliances and Changes in Input Markets

Although the quality of major agricultural crops such as grains and fibers has economic value, they are traded commodities for which market prices and location are more generally determined by competitive markets. In a commercial system, providers and producers of genetic materials for these crops are not expected to become involved in marketing the final commodities. Providers and producers of genetic materials will be interested in obtaining revenues from the sales of the materials, and the farmers who buy the materials will sell the products. Several potential features of the marketing and pricing of genetic material bear mention.

To obtain base varieties that will be modified genetically, biotechnology companies can establish alliances with seed companies or purchase them. Suppose a biotechnology company identifies a genetic manipulation that provides protection from a disease. This disease may afflict a crop many regions, but because of the heterogeneity of agriculture, growers may use many varieties, each adapted to a specific environment. The biotechnology company will seek access to each of these regions by modifying a range of base varieties. The recent purchase of Holden by Monsanto and the alliance between DuPont and Pioneer are examples of this pattern.

If only a subset of available varieties are genetically modified, inter-variety diversity is likely to decrease, as well as the diversity with respect to the modified genes (other factors held constant). As with any seed technology, however, not all farmers are likely to plant all of their crop area to the new, modified variety, because the base variety is not attractive to them. Some farmers may deliberately or inadvertently create genetic "refuges" by planting some land to varieties that have not been modified. Chapter 14 and the chapters in part III explore the supply and demand factors that reinforce or counteract tendencies toward crop genetic uniformity in developing countries.

Prices of genetically modified seed will reflect a charge for the base variety and fees for the added traits. Until the patents that enable genetic engineering of traits expire, fees for these traits will reflect the monopoly power of the patent owners. As discriminatory monopolists, they are likely to recognize differences across locations in the productivity and value of modified traits. Indeed, Monsanto provides a discounted fee for users of *Bt* cotton in North Carolina, because the value of the pest protection it provides is lower there than in other locations (e.g., Texas). The value of pest resistance or other genetically modified traits depends on commodity prices, and trait fees will decrease when agricultural commodity prices decline, especially as government support programs are eliminated. When patents expire, fees will decline, approaching variable production costs if competitive sources of supply emerge.

Finally, various categories of agricultural inputs will become less specialized, and input suppliers are likely to form alliances and provide integrated solutions to agricultural production problems. Greater integration of the pest control and seed industries is already evident. Integration may also evolve at the level of manufacturers and dealers. We may see the emergence of independent consultants who are knowledgeable in designing an optimal production strategy among farmers and several vendors. A third possibility is that some input suppliers will expand their services to include overall management consulting. For example, seed companies may employ consultants to address farmers' pest management, seed selection, and water use problems.

12.4. IMPLICATIONS FOR DEVELOPING COUNTRIES

Agricultural biotechnology has been in the research pipeline for many years, but it only began to generate commercial applications during the second half of the 1990s. Experience with the diffusion of other technologies suggests that, after an initial period of introduction, there will be periods of takeoff, expansion, and broad application of the products of agricultural biotechnology. Much of the development in biotechnology has been in the United States and the developed nations, although biotechnology has been adopted in some developing countries with advanced scientific capability, such as Argentina and China. This chapter has focused on the experiences of developed countries, but some lessons may be gleaned for the research roles and the design of institutions in developing countries.

First, biotechnology provides tools to assure the continuous advances in productivity that are necessary to counteract the effects of a growing human population and environmental change on the agricultural systems of developing countries. While in some cases biotechnology may substitute chemicals and capital, in most it will enhance their productivity. Biotechnology may provide solutions to unsolved pest problems and reduce the use of chemicals that cause environmental and health risks, but it is important to remember that the introduction of the same genes on a large scale can also contribute to the formation of new pathotypes. Policies for encouraging spatial and temporal diversity, such as the establishment of refuges or mosaics, may need to be implemented.

Second, the development of infrastructure to encourage and strengthen linkages between educational and industrial institutions is crucial. Biotechnology is intensive in human capital and its emergence has been concentrated in areas where there is a critical mass of expertise in natural, technological, social, and management sciences, along with access to financial and entrepreneurial resources. Developing countries will need to pool resources and establish such centers. For example, a university may be targeted to excel in biotechnology development and related disciplines, and it may be linked to an industrial park that will attract both domestic and foreign entrepreneurs. The international research centers of the Consultative Group on International Agricultural Research (CGIAR) might be represented in such clusters. Equally important will be the establishment of legal structures and appropriate policies to provide incentives for investment in biotechnology.

Biotechnology companies in developed countries may not invest in forging solutions for problems specific to some developing countries, but, as with any technology, the products of biotechnology will require local adaptation. Consequently, even countries that may not be in the forefront of biotechnology research will need to build related human and industrial capital.

Third, developing countries should encourage the establishment of a private-sector infrastructure for marketing agricultural inputs, including biotechnology. Developed countries rely heavily on networks of private entities to diffuse new technologies. These networks include distributors, dealers, and consultants, whose efforts are complemented by those of public extension and education. Many developing countries maintain unnecessary regulations on the introduction and commercialization of new technologies. Extension services often compete with private companies to provide technologies instead of complementing their services and providing the infrastructure to scrutinize them. The removal of barriers to developing marketing channels for technologies will facilitate the adoption of new products and justify the investment in developing them. As governments in developing countries stop providing farm inputs and seeds, public resources can be directed more toward education and research.

Fourth, developing countries should establish and enforce intellectual property rights for biotechnology and develop institutions of technology transfer. The ability of developing countries to exploit the biotechnology revolution depends on their ability to provide incentives for technology producers in developed countries to export those technologies and invest in the infrastructure that will encourage adoption of their knowledge-intensive products. Suppliers of biotechnology products will lose a significant share of their profits without well-enforced intellectual property rights. Currently, there is debate over the exact definition of intellectual property rights in biotechnology, but over time international agreements will emerge.

Establishing offices of technology transfer at major research institutions in developing countries will enable them to claim rights for genetic materials and license the rights to develop products to companies in both developing and developed nations. Many discoveries of research organizations in developing countries have significant implications and applications beyond the developing world. They can be utilized and funded by commercial farmers throughout the world. A larger base of intellectual property rights in developing countries would give those countries better bargaining positions in collaborative agreements.

Fifth, public-sector research on the genetic enhancement of varieties, especially for subsistence farmers, should continue. Crop breeders should have free or low-cost access to intellectual property rights and other technological components. Subsistence farmers may not provide the commercial incentives necessary to justify genetic enhancement research by private firms. Even when commercial varieties are available, alternatives, such as those provided by public research institutions, will tend to reduce their cost. Public-sector developers of technologies for subsistence farmers may need to use both their economic clout and the political process to reduce the cost of intellectual property rights and obtain cheaper access to genetic materials. On the other hand, agreements that provide for preferential treatment of genetic materials for the poor should be revisited to prevent abuse. For example, biotechnology material that is used in developing countries to produce export goods should not be exempt from paying for intellectual property rights.

Sixth, gene banks should be maintained and developing countries should make efforts to protect and better understand the diversity of their biological and genetic resources. While royalties from biodiversity may sometimes be substantial, the overall economic gains they provide may be limited. Biotechnology provides a means to utilize genetic diversity and increase the value of institutions such as gene banks. The social value of genetic materials in gene banks will increase with greater accessibility and use, but gene banks should consider establishing revenue-generating schemes to support their collection and conservation costs. Public research institutes may obtain materials from gene banks at preferential rates, while private organizations may pay fixed subscription fees that may be augmented by royalties.

Finally, differentiation among the products of agricultural biotechnology poses risks but offers opportunities to developing countries. Some biotechnology research will enable producers in developed countries to grow high-value products that are presently grown in developing countries. For example, algae, soybeans, and canola or other field crops may one day be manipulated to produce silk or other valuable fibers as well as spices, coloring, and fragrances that are important sources of income to some developing countries. Varieties may be modified to withstand weather extremes, reducing the relative advantage of farmers in developing countries.

On the other hand, specific climatic and environmental conditions and low labor costs may encourage multinational and local companies to invest in and facilitate production of differentiated products in certain regions of developing countries. There are already cases in which developing country farmers excel in producing and exporting specialty crops (e.g., the flower sector in Colombia and Ecuador and fresh fruits in Chile and Mexico). There is evidence that contract production of high-value crops is a source of economic development (Carletto, 1996), but it also may lead to negative equity effects and increased economic risks for farmers. Initially the developers of differentiated agricultural biotechnology in developing countries are likely to be multinationals, but they provide the infrastructure (human and physical capital) that may lay the foundation for further growth.

Developing countries should actively pursue opportunities and provide incentives and infrastructure to develop new differentiated industries in agricultural biotechnology. They should also establish standards and procedures for environmental monitoring and evaluation, to avoid possible environmental side effects.

12.5. CONCLUSIONS

After 30 years, agricultural biotechnology has some modest achievements to its credit, but its enormous potential remains to be realized. The experience of developed countries demonstrates that although issues of consumer acceptance and secondary environmental effects may hinder its progress, biotechnology changes the structure of agricultural research. Biotechnology contributes to the privatization of agriculture, creates closer links between universities and industry, encourages product differentiation and contractual arrangements, and can expand the range of agricultural products.

Biotechnology holds special promise for developing countries in their efforts to increase agricultural productivity and reduce resource degradation. To benefit in full from the biotechnology revolution, developing countries will need to improve their agricultural research infrastructure, introduce more efficient distribution networks, provide intellectual property protection, and ensure a hospitable environment for investors and suppliers of technology.

Biotechnology is a promise rather than a panacea. Development of other technologies, such as chemical and mechanical pest control and improved cultural practices, must be pursued, since these technologies will still play a major role in addressing pest problems and improving agricultural productivity. Developing countries should have realistic expectations concerning the value of diversity in their biological and genetic resources; the realization of its potential will be gradual. The challenge is to discover the means to make the right policy adjustments, in the right sequence and at the right time, to reap the benefits of agricultural biotechnology.

Acknowledgments
We gratefully acknowledge research support from the Food and Agriculture Organization of the United Nations, the University of California Systemwide Biotechnology Research and Education Program, and the Giannini Foundation of Agricultural Economics, University of California at Berkeley.

References

Alston, J. M., and P. G. Pardey. 1996. *Making Science Pay: The Economics of Agricultural R&D Policy*. Washington, DC: AEI Press.

Carletto, C. 1996. Non-traditional agro-exports among small holders in Guatemala. Mimeo. Berkeley: Department of Agricultural and Resource Economics, University of California at Berkeley.

Cochrane, W.W. 1993. *The Development of American Agriculture: A Historical Analysis*. St. Paul: University of Minnesota, Department of Applied Economics.

Khanna, M., and D. Zilberman. 1997. Incentives, precision technology and environmental quality. *Ecological Economics* 23(1): 25-43.

Parker, D., D. Zilberman, and F. Castillo. 1998. Office of Technology Transfer, the privatization of university innovations, and agriculture. *Choices* (First Quarter): 19-25.

Postlewait, A., D. Parker, and D. Zilberman. 1993. The advent of biotechnology and technology transfer in agriculture. *Technological Forecasting and Social Change* 43: 271-287.

Schultz, T. W. 1964. *Transforming Traditional Agriculture*. New Haven: Yale University Press.

Williamson, O. 1985. *Markets and Hierarchies: Analysis and Antitrust Implications*. New York: Free Press.

Wright, B. 1985. Commodity market stabilization in farm programs. In *U. S. Agricultural Policies: The 1985 Farm Legislation*. Washington, DC: American Enterprise Institute.

13 ENHANCING THE DIVERSITY OF MODERN GERMPLASM THROUGH THE INTERNATIONAL COORDINATION OF RESEARCH ROLES

G. Traxler and P. L. Pingali

13.1. INTRODUCTION

The scientific ability to exploit genetic resources has been the engine of productivity growth in much of world agriculture for the past 35 years. Developing country wheat yields have risen at 3.4%/yr between 1969 and 1995 (CIMMYT, 1996) and rice yields have risen at an annual rate above 2%/yr (Pingali and Heisey, 1996). Higher input levels and irrigation investments made an important contribution, but the catalyst for the increased productivity has been improved biological technology. The continuing momentum of the green revolution can be explained in great part by the unprecedented cooperation and collaboration between the international agricultural research centers (IARCs) and the national agricultural research systems (NARSs) in the development and exchange of improved germplasm. This cooperation has facilitated the global dissemination of modern, high-yielding germplasm. A second important responsibility of this global research system is to enhance diversity within mandate crops. In this chapter we examine the international research system that supports the dual challenge of increasing agricultural productivity while enhancing genetic diversity.

Prior to 1960, no formal system was in place to provide plant breeders with access to germplasm available beyond their borders. Diversity within individual countries was almost entirely limited to indigenous landraces and varieties derived from combining native germplasm. The present international–national system of cooperation in crop improvement research has increased the number of varieties released in developing countries and the number of landraces represented in released varieties (Smale, 1998). One measure of this success is the increase in temporal diversity. In the late 1960s, 34 wheat varieties and 30 rice varieties were released on average in developing countries each year. By the late 1980s this had increased to 63 wheat and 76 rice varieties.

There is reason to be circumspect, however, when assessing the potential of the international germplasm improvement systems, particularly the wheat system, to sustain the progress it has made in delivering diversity to farmers' fields. The budgets of many international and national research institutions have declined in real terms in recent years. And although defining a precise level of "adequate funding" for research aimed at enhancing genetic diversity is difficult, it seems likely that the wheat and rice systems are underinvesting in these activities at present (Evenson, 1996a). This chapter presents a model of cooperative international crop improvement research and identifies the roles played by institutions participating in the international wheat and rice improvement research systems. The model emphasizes that an efficient and stable international system may comprise partner institutions with a limited breadth of research activities, particularly when research budgets are fixed or declining. Cooperation with complementary institutions, the free exchange of scientific information, and open access to research results are essential for this system to function properly. We examine the experience of this international system over the past three decades and discuss how resources might be reallocated to increase the amount of "upstream" scientific research— that is, research that successfully incorporates the diversity of genetic resources.

13.2. THE INTERNATIONAL RESEARCH SYSTEMS FOR IMPROVING WHEAT AND RICE

The transformation of genetic resources into plant types that are useful for farmers is a cooperative enterprise that links scientists and institutions in virtually all wheat- and rice-growing countries. The development of a new variety involves many steps, from the collection of unimproved landraces and wild species, to storage and characterization, pedigree breeding, and finally the testing of advanced lines in targeted release areas. The entire process may take 30 or more years to complete. The following sections describe the interrelationships of breeding institutions and develop a conceptual model of the link between yield increases and the focus of research in international rice and wheat improvement centers. The model is discussed largely with reference to wheat improvement, but the institutional framework and yield model are similar for rice.

13.2.1. System Participants and Their Research Roles
The international wheat and rice improvement system pursues four classes of germplasm improvement activities: basic research, genetic resource conservation and management, pre-breeding research, and cultivar development (Table 13.1). Scientists do not agree entirely on definitions of each research category or on the boundaries between categories,[1] but loose definitions of the categories will serve to illustrate the simple model presented in the next section of this chapter.

The key activities for promoting diversity are included under genetic resource conservation and management and pre-breeding. Through germplasm bank activities (collecting, characterizing, and screening accessions), new germplasm, including landraces and wild species, enters the crop improvement system and begins to be engineered into a form that farmers can use. Pre-breeding is aimed at separating desirable from undesirable traits in base collections of germplasm and forming elite lines that serve as parent material in developing cultivars targeted to different agroecologies. This process can take 20 years or longer and mostly generates public goods. There are no commercial markets for elite (i.e., unfinished) rice or wheat germplasm, and the benefits from this germplasm enhancement process spill out to be shared by many countries. Cultivar development comprises two main plant breeding activities: crossing lines to create new varieties and evaluating varieties developed elsewhere for possible release. These activities are undertaken with a view to ensuring that improved germplasm, whether developed internationally or nationally, is targeted closely to particular agroecologies where wheat (or rice) is grown. The benefits of this kind of research are captured by the countries that conduct it. International screening and testing of advanced breeding materials in trials coordinated by the IARCs is an essential

Table 13.1. Wheat Crop Improvement Research Activities and Institutions

Wheat improvement activities	International research institutions and roles
I. Basic research	**Developed country institutions** (90 scientists[a])
	Basic research
II. Genetic resource management	Genetic resource management
A. Acquiring and maintaining germplasm	Pre-breeding research
B. Documenting accessions	
C. Biotic and abiotic stress screening	**International Maize and Wheat Improvement Center (CIMMYT)** (30 scientists)
	Genetic resource management
III. Pre-breeding research	Pre-breeding research
A. Inter-generic and inter-specific crossing	Cultivar development
B. New plant types/yield frontier research	
C. Introgressing genes for new disease resistance	**Stage 3 NARSs** (658 scientists)
D. Administering screening nurseries	Pre-breeding research
E. Administering yield nurseries	Cultivar development
IV. Cultivar development	**Stage 2 NARSs** (87 scientists)
A. Pedigree crossing/breeding program	Adaptive breeding program using CIMMYT parents
1. Breeding pool improvement	Testing and screening cultivars from outside sources
2. Development of advanced lines	
3. Crossing of own lines	**Stage 1 NARSs** (409 scientists)
B. Adaptive breeding program	Testing and screening cultivars from outside sources
1. Crossing or selecting lines from outside sources	
2. Screening cultivars from outside sources	
C. Testing/screening program (testing cultivars developed elsewhere)	

[a] Researchers in US and Australia only. Australian scientists assumed to be allocated among research foci in same ratio as reported for the US. Sources for numbers of scientists: Frey (1996), Byerlee and Moya (1993).

component of cultivar development. This international flow of wheat and rice germplasm has had an enormous impact on the speed and the cost of cultivar development.

Three kinds of research institutions[2] participate in the international wheat and rice improvement systems. These are public institutions (universities and other institutions) in developed countries; IARCs (for wheat these are the International Maize and Wheat Improvement Center—CIMMYT, and the International Center for Research in the Dry Areas–ICARDA) and the NARSs of developing countries (Table 13.1). The 31 NARSs can be further subdivided by the breadth and output of their research programs (Byerlee and Traxler, 1995) into Stage 1, 2, and 3 NARSs. Stage 3 NARSs do some pre-breeding research to enhance the diversity of their breeding material and develop a significant amount of parent material for their crossing programs (Table 13.2). At present Brazil, India, China, and Argentina demonstrate Stage 3 breeding capacity for wheat, while

Table 13.2. Number of Wheat and Rice Varieties Released from Own Crossing Program, and Release of Screened IARC Varieties, by Country, 1977–1990 (for countries producing 10 or more varieties from own crosses)

Crop/country	NARS cross	NARS screened	Total releases
Wheat			
Stage 3 NARSs			
Brazil[a]	84	34	118
India[a]	83	28	111
Argentina	34	35	69
Southern China	26	10	36
Stage 2 NARSs			
Chile	14	31	45
Kenya	13	3	16
Stage 1 NARSs			
Remaining 25 NARSs	61	275	336
Total	**315**	**416**	**731**
Rice			
Stage 3 NARSs			
India[a]	251	36	287
Korea	88	0	88
Brazil	66	7	73
China	51	2	53
Sri Lanka	32	0	32
Burma	32	16	48
Nigeria	27	4	31
Ivory Coast	23	0	23
Indonesia	20	13	33
Thailand	16	0	16
Stage 2 NARSs			
Malaysia	14	2	16
Philippines	13	20	33
Mexico	10	2	12
Stage 1 NARSs			
Remaining 40 NARSs	120	85	205
Total	**763**	**187**	**950**

[a] "Super NARSs."

ten countries do so for rice. Stage 2 NARSs have crossing programs that have demonstrated a minimum level of success in cultivar development, producing about one new variety every year from their own crosses. The Stage 2 NARSs in wheat are Chile and Kenya; in rice they are Malaysia, the Philippines, and Mexico. The remaining NARSs, which include 25 national wheat programs and 40 national rice programs, do not have effective crossing programs. These NARSs release useful varieties by testing imported varieties and releasing those best adapted to their target environments.[3] Stage 2 and 3 NARSs do not engage in pre-breeding, but they may contribute to genetic diversity by increasing the rate of varietal turnover.

Developed country institutions are the major source of basic research, because basic research easily spills across geographic and crop species boundaries (e.g., between winter and spring wheat environments). Their genetic enhancement research has also had very large impacts in developing countries, but on a more sporadic basis,[4] since much of their research is not relevant for agroclimatic conditions in developing countries. CIMMYT is largely a borrower of basic research, but it is the sole important source of international spillovers of genetic resource conservation and management, pre-breeding research, and cultivar development research. In 1990, 42% of the spring wheat area in developing countries was planted to CIMMYT varieties and another 24% of the area was planted to varieties that had CIMMYT parents (Byerlee and Traxler, 1995). Only 15% of the area worldwide was planted to modern varieties that did not have CIMMYT parents, and 90% of that area was in India.[5] Even in India, CIMMYT materials are evident in the pedigrees of the wheats produced. To date, NARS–to–NARS exchanges have not been an important source either of finished varieties, except in occasional cases such as India and Nepal, or of pre-breeding material.

International screening and testing of germplasm enormously reduced the costs of adaptive breeding for NARSs. The efficiency of cultivar development has been enhanced by the existence of such a network.[6] Capitalizing on the network, NARSs have achieved the goal of improved wheat productivity. The most successful national wheat breeding scientists focus on screening and testing imported cultivars (Maredia and Byerlee, 1998). Even countries with successful breeding programs rely heavily on imported cultivars for much of their pre-breeding material.

13.2.2. Balancing Diversity-enhancing and Cultivar Development Research

The international wheat and rice improvement systems' continued success in producing improved varieties for farmers will require "sufficient" levels of all four classes of research. In the long run, yields cannot be improved unless pre-breeding material with higher yield potential is available. National research programs' ability to maintain and enhance genetic diversity is even more dependent on access to the results of pre-breeding research. In the short run, however, yield improvement research and research aimed at increasing diversity may compete for resources. In this section we discuss a simple model of the trade-off implicit in balancing genetic enhancement research and cultivar development research.

Precise information on the actual distribution of plant breeders within the international rice or wheat system is not available, but the evidence for wheat suggests that the balance is tipped heavily toward cultivar development research, with a potentially serious lack of support for genetic resource management and pre-breeding research. Using the number of scientists reported in Figure 13.1, a rough estimate of the total number of scientists who conduct genetic resources and pre-breeding research would represent 7% of the scientists in the CIMMYT/national agricultural research system. This is based on the assumption that the efforts of CIMMYT's wheat scientists are equally divided between genetic enhancement and cultivar development research, and that the efforts of 10% of Stage 3 NARS scientists are dedicated to genetic enhancement research. If the 7% estimate is correct, the international wheat system dedicates less than half of the effort to genetic enhancement that is allocated in the US (Frey, 1996). Evenson (1996a) suggests that the international rice improvement system also seriously underinvests in genetic enhancement research.

A number of factors explain the skewed pattern of research investments toward cultivar development. The full benefits from investments in research to enhance genetic resources are difficult for any single country to capture. They also occur with long lags, are highly uncertain, and require a more sophisticated scientific capacity. Countries with small growing areas behave rationally by choosing to free ride on the international system rather than to invest in a large wheat breeding infrastructure (Maredia, Byerlee, and Eicher, 1994). As long as the international public good is available, such decisions enhance the short-run efficiency of NARSs. Such NARSs may view diversity-enhancing research as a luxury to be funded only after more immediate, and more readily appropriated, cultivar development research needs are met. Because IARCs are the only institutions with worldwide mandates, logically they are the institutions to compensate for any lack of conservation and pre-breeding research in the national systems by developing new and increasingly diverse germplasm that NARSs can use to develop new cultivars.

The IARCs provide several types of inputs to the cultivar development programs of NARSs. The first type is broadly adapted, finished varieties. The national systems test these varieties, renaming and releasing those that perform best in their particular

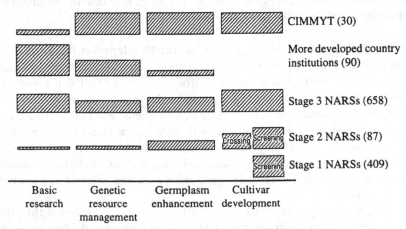

Figure 13.1. Focus of Wheat Improvement Research Institutions
Note: Number of wheat breeders in each class of institution is in parentheses.

environment. Germplasm from IARCs is also used as parent material in NARSs' crossing programs. In Figure 13.2, a production function is used to model the relationship between IARC and NARS crop improvement research. The figure attempts to demonstrate the effect of moving IARC research upstream—that is, of reallocating resources away from cultivar development and toward pre-breeding activities that enhance genetic diversity.

In Figure 13.2, yield in the NARS home environment is drawn on the Y axis. When the IARC conducts a significant amount of cultivar development research, a yield level such as Y_C will be attained by the IARC variety that performs best in the NARS environment. The NARS operates on production function, $Y(S)_{NARS}$, which is an increasing function of the size of its program.[7] As drawn, only a crossing program larger than size S_{min} will succeed in producing varieties from its own crosses that outperform broadly adapted IARC varieties. The research production function eventually flattens out to the point where employing additional scientists does not increase yields beyond Y_{max}. In other words, a maximum yield potential or "yield plateau" is eventually reached, given the existing stock of research output from pre-breeding activities.

What would be the effect of reallocating IARC resources to conduct more research oriented toward enhancing genetic diversity? With a fixed total research budget, fewer broadly adapted, finished varieties will be released to NARSs. This will imply a reduction in the yield of the unadapted IARC variety but will increase the yield potential of *adapted varieties*. Fewer finished varieties will be produced by the IARC, so the expected yield performance of IARC varieties will decline. The yield of IARC varieties will shift down from Y_C to Y_C'. But because of the IARCs' increased genetic enhancement research, NARSs will have better parent material to work with. This changes the shape of the NARS research production function to $Y(S)'$ from $Y(S)$. Although the new IARC parent material increases the yield potential from Y_{max} to Y_{max}', NARSs will be required to increase their investment in cultivar development to realize the yield potential in their home environment.

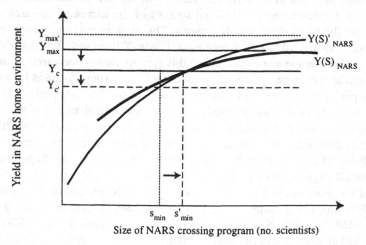

Figure 13.2. Output of NARS Crossing Program with an Increase in IARC Diversity Research.

13.3. SOURCES OF IMPROVED VARIETIES OF WHEAT AND RICE

How is the research effort currently shared between NARSs and IARCs? Given the large investments in NARSs over the past three decades, are NARSs now assuming a larger burden of cultivar development, thereby freeing IARCs to re-deploy their resources to activities that bring new germplasm into the system? To examine these questions, we review data on releases of improved varieties of wheat and rice.[8] The wheat release data cover the 1,248 wheat varieties released by NARSs in 31 countries from 1966 to 1989. The rice data cover 1,540 varieties released by NARSs in 61 countries over the same period.

Neither CIMMYT nor the International Rice Research Institute (IRRI) releases varieties for direct use by farmers. Rather, each institution makes advanced lines available to NARS screening programs. The national research systems may choose to rename and release these varieties after two to three years of testing in their countries. Breeding lines from IARCs may also be entered into NARSs' breeding collections to be used as parent material in national crossing programs.[9] In the analysis that follows, we refer to the output of NARS testing programs (i.e., programs to evaluate varieties developed from crosses made by an IARC) as "screened CIMMYT" or "screened IRRI" varieties. Varieties released from NARS crosses are divided into three groups: (1) those released by "Super-NARSs" (India for rice; India and Brazil for wheat); (2) those released by other NARSs; and (3) NARS–NARS transfers, or third-country releases. The number of varieties in the last category is an indicator of the level of regional cooperation in plant breeding.

Recall that both the rice and the wheat research systems have been great successes in terms of the increased number of varieties being made available to farmers. Even during the post-green revolution years since 1975, the total number of global releases for each crop from all sources (including both NARS and IARC crosses) has continued to increase. Between 1975–79 and 1985–89, the average number of wheat varieties released worldwide each year increased from 50 to 72, and rice releases increased from 74 to 83. However, the institutional sources of the increased research output differed across the two crop research systems. The main increase in wheat releases was the result of increased NARS screening of CIMMYT crosses. Between 1975–79 and 1985–89, the average annual release of this type of variety increased from 24 to 39. Slight increases occurred in the output of NARSs' crossing programs (from 23 to 27 varieties per year), and of third-country releases (from 2 to 6 varieties per year). Super-NARSs' releases of wheat varieties were constant through time. For rice, global output increased primarily because of increases in NARS crosses, which rose from 30 to 40 varieties per year. The number of screened IRRI varieties that were released actually fell by 50%, from 16 to 8, between 1975–79 and 1985–89. Super-NARSs' releases also fell slightly, while third-country releases remained constant.

Some information on more intensive breeding of IARC material is also available. Nearly 70% of the wheat varieties released from NARS crosses in 1986–90 had at least one CIMMYT parent. Nearly 20% of rice varieties released from NARS crosses in 1986–91 had an IRRI parent (Evenson, 1996b), but many more rice varieties had an IRRI grandparent.

Where do these changes leave the world rice and wheat improvement systems as we enter the 21st century? One perspective is provided by reviewing the shares of global varietal releases (Figures 13.3 and 13.4). These data reveal several things about how varietal improvement is currently organized. First, the rice and wheat systems differ greatly in the prominence of IARC crosses as a source of varietal releases. In wheat, more than half of all releases in 1985–89 consisted of CIMMYT crosses screened by NARSs, and the share of releases in this category has increased over time. In fact, there is *no country* in which screened CIMMYT varieties are not an important source of farm-level technology. The number of wheat releases each year from NARS crossing programs has increased slightly over time, but their share in total releases has declined significantly. On the other hand, IRRI is no longer an important source of finished rice varieties. Screened IRRI crosses accounted for only 9% of rice releases in 1985–89, while NARS crosses accounted for 85% of releases worldwide. Scientific resources at IRRI have been freed to engage in work that increases genetic diversity.

Another finding that is relevant to the future of the international wheat and rice research systems is that NARS–NARS transfers of germplasm are not an important component of either system. Discussion of supplanting the IARC system with regional cooperative systems would appear to be premature; moreover, there has been little discussion to date of the complex institutional issues related to funding and operating such regional systems. Regardless of the potential merit of moving to regional rather than—or more likely, in addition to—global research systems, such a move may be decades off.

The fact that IRRI is no longer an important source of finished rice varieties has allowed the institute to divest itself of producing finished varieties, turning responsibility for cultivar development over to the NARSs' crossing programs. Evenson (1996b) presents evidence that IRRI now plays a central role in rice genetic enhancement research, incorporating landraces into breeding material and serving as an essential

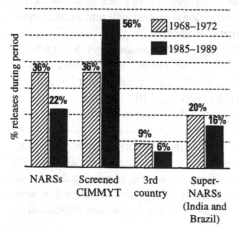

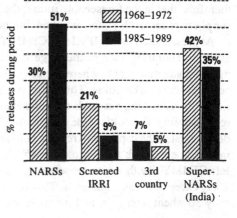

Figure 13.3. Shares of Wheat Releases by Source of Crosses.

Figure 13.4. Shares of Rice Releases by Source of Crosses.

conduit for the global distribution of landrace genes. Some 70% of all landraces that have appeared in the background of NARSs' rice varieties have entered through the use of IRRI crosses. Had the crossing programs of national research systems been less successful, IRRI would not have been able to invest as heavily in its present diversity-increasing research efforts.

CIMMYT, on the other hand, faces a stark dilemma. It is the only important international supplier of upstream research aimed at enhancing genetic diversity, and there appears to be substantial underinvestment in this activity worldwide. Yet with so few NARSs having demonstrated the ability to develop varieties from their own crosses, it would be risky for CIMMYT to increase research that generates new and novel sources of genetic diversity and to reduce efforts to produce varieties for national screening programs. This commitment to both kinds of research strains CIMMYT's resources. The implied mandate is to conduct more upstream research without reducing activities related to cultivar development, within a stagnant budget for wheat breeding. The fact that NARS research budgets are also mostly stagnant or declining suggests that national programs will be hard pressed to improve their performance in cultivar development.

13.4. SUMMARY AND CONCLUSIONS

The international community faces the challenge of devising a system of crop improvement research that supplies germplasm with ever-increasing yield potential, while at the same time devoting adequate resources to longer term concerns for genetic diversity and genetic resource conservation. Insufficient resources are presently devoted to diversity enhancement research, and most of this research is done by IARCs. How can the system increase the resources devoted to incorporating new and novel sources of genetic diversity into germplasm at a time when both international and national research institutions face stagnant real budgets? One option would be to increase regional research coordination, but no such regional system is in place. It seems likely, therefore, that in the foreseeable future, the international centers will remain the major suppliers of research aimed at enhancing genetic diversity.

At present the roles of IRRI and CIMMYT differ in fundamental ways. Most wheat-growing countries, except the large ones, demand improved varieties that require minimal adaptive work, whereas rice-growing countries have far greater capacity to conduct adaptive breeding. The "IRRI model" has permitted IRRI to shift to upstream research, increasing research that enhances genetic diversity and divesting itself of responsibility for cultivar development. By contrast, relatively few NARSs have demonstrated the capacity to release cultivars at a rate that supports temporal diversity in wheat, even with access to CIMMYT's broadly adapted parent material. This has left CIMMYT with major responsibilities for all research areas represented in Table 13.1, except for basic research. This research burden is large for an institution with only 30 wheat scientists, and it is reasonable to ask whether the present allocation of research responsibilities is sustainable. In a world of shrinking budgets, the international wheat system faces a difficult choice. The supply of nearly finished varieties comes at the cost of diverting resources from pre-breeding and other research to enhance genetic diversity. Without adequate levels of investment in pre-breeding, the capacity of the international research system to provide the germplasm on which future economic gains depend is limited.

Acknowledgments

Helpful comments from Sanjaya Rajaram, Tony Fischer, Michael Morris, and Paul Heisey are gratefully acknowledged. The authors also thank Douglas Gollin and Robert Evenson for providing data on the release of rice varieties.

Notes

1 For example, it might be argued that "strategic research" is a more relevant concept than "basic research" for applied research institutions (Tony Fischer, personal communication).
2 The private sector also does a small amount of research but only in a few countries, and it is not a major provider of wheat improvement research in any developing country. This situation may change if laws protecting intellectual property in pure line crops are strengthened. In the United States, for example, 41% of wheat breeding scientists and 52% of rice breeding scientists are employed by the private sector (Frey, 1996).
3 Stage 1 countries focus entirely on screening varieties from international centers. Stage 2 countries screen IARC varieties but also develop a few varieties from adaptive crosses of IARC lines. Stage 3 countries also screen and adapt IARC lines, but they develop a significant number of varieties that have no IARC parents. The majority of wheat releases in all countries except Brazil contain a CIMMYT parent. Worldwide, 85% of the spring wheat varieties that have been released were created using CIMMYT lines, either in screening or adaptive breeding programs. See Byerlee and Traxler (1995) for more detail on wheat.
4 For example, N.E. Borlaug created the first high-yielding (green revolution) varieties by introgressing the Japanese Norin 10 dwarfing gene into his breeding lines through wheat germplasm acquired from Washington State University (Dalrymple, 1986). Research coordinated by Oregon State University and CIMMYT was vital to developing the most successful cross of the 1980s, the Veery lines (which were the product of a winter × spring wheat cross).
5 The remaining 19% of the wheat area in developing countries was still planted to tall varieties. All area figures exclude China.
6 Maredia and Byerlee (1998) provided evidence that the efficiency of many national breeding programs would be improved if they reduced the size of their crossing programs. Table 13.2 shows that only a handful of NARS crossing programs have been successful in releasing varieties.
7 Byerlee and Traxler (1996) provided evidence that the output of crop improvement research programs increases with program size.
8 Wheat release data were collected by CIMMYT as part of a 1990 impact study; results of that study are presented in Byerlee and Moya (1993). Data on rice releases were collected by T. R. Hargrove and V. Cabanilla of IRRI.
9 Information on the use of IRRI material as parent lines is not available. More than half of NARSs' wheat releases in the 1980s had a CIMMYT parent.

References

Byerlee, D., and P. Moya. 1993. *Impacts of International Wheat Breeding Research in the Developing World, 1966–1990*. Mexico, D.F.: International Maize and Wheat Improvement Center (CIMMYT).
Byerlee, D., and G. Traxler. 1995. National and international wheat improvement research in the post-green revolution period: Evolution and impacts. *American Journal of Agricultural Economics* 77: 268–78.

Byerlee, D., and G. Traxler. 1996. The role of technology spill-overs and economies of size in the efficient design of agricultural research systems. Paper presented at the conference on Global Agricultural Science Policy for the Twenty-First Century, 26–28 August, Melbourne, Australia.

Dalrymple, D. G. 1986. *Development and Spread of High Yielding Wheat Varieties in Developing Countries*. Washington, DC: US Agency for International Development.

Evenson, R. E. 1996a. Valuing agricultural biodiversity. Paper presented at the conference on Global Agricultural Science Policy for the Twenty-First Century, 26–28 August, Melbourne, Australia.

Evenson, R. E. 1996b. Rice varietal improvement and international exchange of rice germplasm. Paper presented at the Conference on the Impact of Rice Research, 3–5 June, Bangkok, Thailand.

Frey, K. 1996. *National Plant Breeding Study - I: Human and Financial Resources Devoted to Plant Breeding Research and Development in the United States in 1994*. Special Report 98. Ames: Iowa Agricultural and Home Economics Experiment Station.

Maredia, M. K., and D. Byerlee. 1998. *Research Efficiency in the Presence of Technology Spillovers: The Case of National and International Wheat Improvement Research*. CIMMYT Research Report. Mexico, D.F.: International Maize and Wheat Improvement Center (CIMMYT).

Maredia, M. K., D. Byerlee, and C. K. Eicher. 1994. *The Efficiency of Global Wheat Research Investments: Implications for Research Evaluation, Research Managers, and Donors*. Staff Paper No. 94-17. East Lansing, Michigan: Michigan State University.

Pingali, P. L., and P. W. Heisey. 1996. Cereal crop productivity in developing countries: Past trends and future prospects. Paper presented at the conference on Global Agricultural Science Policy for the Twenty-First Century, 26–28 August, Melbourne, Australia.

Smale, M. 1988. Indicators of genetic diversity in bread wheats: Selected evidence on cultivars grown in developing countries. In R. E. Evenson, D. Gollin, and V. Santaniello (eds.), *Agricultural Values of Plant Genetic Resources*. Wallingford: CAB International.

14 ACHIEVING DESIRABLE LEVELS OF CROP DIVERSITY IN FARMERS' FIELDS: FACTORS AFFECTING THE PRODUCTION AND USE OF COMMERCIAL SEED

M. L. Morris and P. W. Heisey

14.1. INTRODUCTION

Based on theoretical considerations alone, it may soon be possible to determine the level of crop diversity that could be considered socially optimal for a particular geographical area. A productivity-based definition of the social optimum might be the level of diversity that maximizes expected aggregate productivity across a specified region, subject to a minimum acceptable level of protection against catastrophic crop losses from pests and/or diseases. A conservation-based definition would reflect other criteria, such as allelic diversity in a reference region.

In either case, once it becomes possible to define genetic diversity (in a theoretical sense) and to measure it (in a practical sense), policymakers will face a problem. What combination of social and/or economic institutions, as well as incentives, will be needed to ensure that the socially optimal level of diversity is actually achieved? The problem is hardly trivial, because utility-maximizing behavior on the part of farmers and seed companies could conceivably lead to widespread adoption of a limited number of varieties. Depending on the nature of the relationship between spatial and temporal distributions of varieties and genetic diversity, this could result in an undesirable decrease in crop genetic diversity measured across broad geographical areas.

This chapter examines the factors that influence the adoption and diffusion of seed of modern varieties. Adoption and diffusion are two parts of a process that directly affects the level of crop genetic diversity found in farmers' fields. The chapter begins with a review of the empirical evidence on the diffusion of modern varieties of rice, wheat, and maize. The meaning of the term "modern variety" differs by crop, as explained later in the text. The next section examines factors that affect the demand for seed of modern varieties. We then discuss developmental patterns in national seed industries and analyze the technical, economic, and institutional factors that influence the supply of crop seed. Although this section focuses on formal seed systems, the relationship of informal and formal systems, and the contribution of information systems to crop diversity, are noted briefly. The final section of the chapter examines policy options for influencing the supply and demand of commercial seed and spells out implications for crop genetic diversity at the farm level. The analysis leads us to conclude that neither farmers nor seed companies are likely to have strong incentives for increasing crop diversity; therefore, policies aimed at increasing crop diversity at the farm level will probably be difficult to design and implement.

The focus throughout the chapter remains primarily on commercial agriculture, i.e., agriculture in which input use decisions are driven by economic considerations that are mediated by the market. The analysis holds less relevance for cropping systems in which farmers make limited use of purchased external inputs, including seed of modern varieties. Similarly, the discussion of policy implications is most pertinent to the commercial cropping sector, which responds most directly to the types of policy measures described in the concluding section. Consistent with the rest of this book, our empirical examples concentrate on the three major cereals. Wheat, rice, and maize currently account for approximately 85% of total world cereal crop production. Although not discussed here, work on informal seed systems (Almekinders, Louwaars, and de Bruijn, 1994) is also relevant to the crop diversity discussion. Informal channels are especially important for crops like beans for which home consumption characteristics are particularly valued (Sperling, Scheidegger, and Buruchara, 1996). Farmer–to–farmer exchange can also be a major conduit for seed of self-pollinated cereal crops such as wheat, even when all farmers are growing modern varieties (Heisey, 1990).

14.2. DEFINING AND VALUING CROP GENETIC DIVERSITY

Before we can speak meaningfully of a socially optimal level of crop genetic diversity, it is necessary to have a workable definition of diversity, as well as some practical means of measuring and valuing it. Various indicators have been proposed to measure genetic diversity in crop plants. As discussed in chapter 2, these indicators often differ in their level of analysis and conceptual basis. Across a specified set of farmers' fields located within a specified area (e.g., district, state, country, region), attention is usually focused on the spatial and temporal distribution of varieties (Souza *et al.*, 1994; Brennan and Byerlee, 1991; Duvick, 1984). Occasionally, spatial and temporal measures are combined with pedigree-based measures. At present, however, the cost of using molecular data on such a wide scale is prohibitive. In addition, as mentioned in chapter

2 and suggested by the studies in part IV of this volume, it is difficult to establish a relationship between either molecular or pedigree-based measures and field performance over a geographical expanse.

How can crop genetic diversity be valued in economic terms? In the abstract, it may be assumed that society directly or indirectly derives varying amounts of utility from different levels of genetic diversity; this relationship can be used to trace out a utility function. If more than one diversity indicator is valid, a diversity frontier can be constructed indicating the highest level attainable of a particular indicator, given fixed levels of the other indicators; more generally, a mixed goods–diversity frontier can be constructed. The point of tangency between this frontier and the iso-utility curve indicating society's preferences indicates the socially optimal level of genetic diversity. Since attaining any given level of genetic diversity involves a policy cost, in any particular situation the socially optimal level will depend on the point at which marginal benefits equal marginal costs.

It is important to note two points. First, no matter how sophisticated the measure of genetic diversity, and no matter how complicated the representation of society's preferences, it must be possible to map these abstract concepts on to spatial and temporal distributions of varieties in order for them to have policy relevance. Second, this implies that factors affecting the supply and demand of varieties, and particularly the seed that is the delivery vehicle for these varieties, will be of central importance.

14.3. ADOPTION AND DIFFUSION OF MODERN VARIETIES

Before considering the extent to which farmers currently grow modern varieties, we should first clarify what we mean by the term "modern varieties" (MVs). For wheat and rice, a fairly restrictive definition of MVs is used here; in most areas, the estimates refer specifically to semidwarf varieties. Primarily because of data limitations, we have chosen not to use the broader definition of MVs used elsewhere in this book ("any product of a scientific crop improvement program").

For maize, the definitional problem is different. Because of the way maize reproduces, unless the process is carefully controlled (as it is, for example, in commercial seed production plots), individual plants may be self-pollinated (fertilized by pollen from the same plant), cross-pollinated (fertilized by pollen from other plants growing in the same field or in neighboring fields), or both. As a result, even plants of the same variety are genetically unique.[1] Furthermore, the genetic composition changes from generation to generation. This greatly complicates the matter of defining "modern" or "improved" germplasm. When a scientifically bred, open-pollinated variety (OPV) is maintained at the farm level through repeated recycling of farmer-saved seed, at what point do genetic changes caused by outcrossing alter the variety to the extent that it can no longer be considered "improved"? Since there is no theoretical basis for answering this question, any definition of what constitutes an "improved OPV" is basically arbitrary. Here we have tried to follow the rule-of-thumb used in CIMMYT's studies of the impact of maize research, in which we classify as "modern" any crop grown using seed from a commercial source or farmer-saved seed that has been recycled no more than twice since it was originally purchased from a commercial source.[2]

14.3.1. Area Planted to Modern Cereal Varieties

Table 14.1 summarizes information on the area planted to MVs of wheat, rice, and maize worldwide. In industrialized countries, although not all rice and wheat area is planted to semidwarfs, virtually all of the area planted to rice, wheat, and maize is sown to germplasm derived from scientific plant breeding efforts. In developing countries, the figures are lower, ranging from around 74% for wheat and rice to around 60% for maize.[3]

14.3.2. Temporal Diversity

The adoption data in Table 14.1 provide a useful snapshot of MV use during the mid-1990s, but this picture is static. In reality, MV adoption and diffusion are dynamic, as farmers continually abandon older varieties in favor of newer ones. In areas first affected by the so-called "green revolutions" in rice and wheat (e.g., northwestern Mexico, southern Turkey, the Punjabs of India and Pakistan, central Luzon in the Philippines, the island of Java in Indonesia, southern China), farmers have replaced their MVs at least twice since they originally adopted semidwarf varieties (Byerlee, 1996).

What do we know about rates of change in varieties? The most comprehensive evidence for varietal turnover in major cereals is available for wheat. Brennan and Byerlee (1991) developed and applied a measure of the temporal diversity of wheat varieties, "weighted average varietal age," calculated as the average age of all varieties grown in a certain region in a specified year, weighted by the area planted to each. The weighted average ages of wheat varieties have been reported for countries and for specific production zones in the developing and industrialized worlds (Brennan and Byerlee, 1991), for countries in the developing world (Byerlee and Moya, 1993), and for countries and larger geographical regions in the developing world (Smale, 1996).

The weighted average ages calculated by Brennan and Byerlee for 1986 ranged

Table 14.1. Percent Area Planted to Modern Varieties of Wheat, Rice, and Maize

	Wheat		Rice		Maize		
	Area planted to semidwarf wheat varieties[a] (%)	Year of estimate	Area planted to modern rice varieties[a] (%)	Year of estimate	Area planted to maize hybrids (%)	Area planted to improved maize OPVs (%)	Year of estimate
Sub-Saharan Africa[b]	60	1994	15	1983	36	14	1992
West Asia/North Africa	42	1990	11	1983	22	7	1992
Asia (excluding China)	91	1994	67	1991	10	29	1992
China	70	1994	100[d]	1991	90	7	1992
Latin America[c]	92	1994	58	1991	37	13	1992
All developing countries	74	1994	74	1991	45	15	1992
Industrialized countries	55	1994	78	1991	99	0	1992
World	64	1994	74	1991	63	10	1992

Source: Byerlee (1996); CIMMYT (1993, 1994, 1996); IRRI (1995); Pardey *et al.* (1996).
a Excludes scientifically bred tall varieties.
b Includes South Africa.
c Includes Brazil, Argentina, and Chile.
d Over half of China's rice area is currently planted to hybrid rice.

from a low of 3.1 years in the Yaqui Valley of Mexico to a high of 11.1 years in the Punjab of Pakistan, with a mean of slightly more than 7 years. Generalizing across all these empirical studies, it is clear that results are sensitive to the selection of countries and the scale of the analysis. Within large countries, the turnover of varieties can vary markedly from region to region. The overall weighted average age of wheat varieties in India is about 12.5 years, for example, but the same figures measured at the state level range from 6–8 years for Punjab to well over 16 years for Bihar (Byerlee and Moya, 1993; Smale, 1996).

As far as we know, weighted average ages have never been calculated for rice varieties, although references to rates of turnover in rice appear in the literature with some frequency. For example, Pardey *et al.* (1992) and Jatileksono and Otsuka (1993) report that farmers in Indonesia replace modern rice varieties every 10 years or less. In chapter 6, Bellon *et al.* report that in the Philippines, farmers in three ecosystems have been planting modern rice varieties for an average of between 2.7 and 3.8 years. These figures suggest weighted average ages for modern rice varieties of 5–10 years. In contrast, farmers in upland and rainfed lowland ecosystems of the Philippines have planted traditional rice varieties for much longer periods, with means ranging from 9 to 27 years. This suggests that weighted average ages for traditional rice varieties were much larger.

López-Pereira and Morris (1994) calculated weighted average ages for maize varieties for 1990. Among improved OPVs, these averaged 10 years for the sample as a whole, with most of the large maize-producing countries falling between 5 and 15 years. In countries where maize hybrids are relatively well established, the aggregate weighted average age for hybrids was 11 years, ranging from a low of 3 years in Brazil to over 20 years in India and El Salvador. Similar figures are not available for industrialized countries, but Duvick (1992; 1998) stresses that in the United States newer hybrids are constantly replacing older ones.

On the whole, the evidence suggests that the rate of varietal turnover varies tremendously between crops and between countries. Although it is risky to generalize, two broad patterns can be discerned. First, in developing countries the rate of turnover seems to have been greater for wheat and rice varieties than for maize varieties. It is important to remember, however, that genetic differences among these crops mean that intra-varietal change is expected to be more rapid for open-pollinated maize varieties than for wheat or rice (see chapter 2 for a definition of intra-varietal diversity). Second, the rate of turnover seems to have been greater in areas dominated by commercial cropping systems than in areas that feature large numbers of subsistence-oriented farmers.

14.3.3. Latent Diversity Measured with Pedigrees
How does the turnover of varieties in farmers' fields affect crop genetic diversity? Since we are interested in assessing how genetic diversity may have changed with the diffusion of scientifically bred crop varieties, we will review the evidence relating to latent diversity among MVs, as measured by pedigrees. In general, more evidence is available for wheat and rice than for maize.

By some measures, the diversity of scientifically bred rice and wheat varieties has increased over time. During the past 30 years, modern bread wheat varieties released in developing countries have benefited from more sustained selection at the hands of formally trained scientific breeders, have featured greater numbers of landraces in their pedigrees, and have become increasingly complex, in the sense that they have been developed using greater numbers of parents (Smale, 1996). Trends in rice breeding have been similar. For example, the number of landraces appearing in the pedigrees of modern rice varieties has increased steadily, and the total number of ancestors incorporated in released MVs appears to have grown by 25–50 almost every year since 1970 (Evenson and Gollin, 1997). A "typical" modern variety of spring bread wheat released in developing countries today may be derived from 45–50 landraces; a typical modern variety of rice from 25 or more.

Less evidence is available on trends in the diversity of modern maize varieties, in part because the pedigrees of most commercial maize hybrids are trade secrets and cannot be analyzed. Circumstantial evidence suggests that diversity in maize is increasing, however. In a recent survey of maize breeding activities in developing countries, López-Pereira and Morris (1994) reported that as national breeding programs grow stronger, breeders stop simply importing and releasing foreign materials and initiate increasingly sophisticated local crossing programs. A general broadening of the genetic base has also occurred in crops in industrialized countries; surveyed in 1981, breeders of five crops (wheat, maize, sorghum, cotton, soybean) in the United States indicated that most had broadened their breeding base considerably in the previous ten years (Duvick, 1984).

Despite the trend toward greater diversification in the pedigrees of MVs released over the past decades, there are strong reasons for varieties within any given adaptation zone to have a high degree of correlation in their pedigrees (Duvick, 1984). Table 14.2 presents estimates of the latent diversity of advanced breeding materials or varieties of wheat, rice, and maize, as measured by pedigree analysis (see chapter 2 for explanations of pedigree analysis and latent diversity). Pedigree-based measures of latent diversity have been widely used, in part because of their relative ease of computation. Furthermore, despite their limitations, they allow us to make some general observations regarding patterns in crop genetic diversity.

Judging from the data presented in Table 14.2, there appears to be no systematic relationship between levels of latent diversity and crop species. Nor is there any obvious relationship between levels of latent diversity and the degree of economic development of the country in which the crop is grown. Not surprisingly, levels of diversity tend to be lower the more disaggregated the analysis and the more homogeneous the production region (in other words, levels of diversity tend to be lower within narrowly defined production zones than across entire countries or regions). With a few exceptions, time trends usually do not show either sharp increases or sharp reductions in latent diversity; rather, to the extent that they change at all, levels tend to drift slowly over time. In some cases (e.g., wheat in the Yaqui Valley of Mexico), high temporal diversity, resulting from rapid turnover in varieties, appears to have substituted for low spatial diversity. In the Canadian Western Red Spring Wheat pool, the rate of turnover and spatial diversity have both been very low.

14.3.4. Spatial Diversity

The measures of latent diversity presented in Table 14.2 relate to the diversity found within sets of released varieties. But to what extent is diversity measurable in the crops actually planted in farmers' fields? Most of the studies that have addressed this question have analyzed patterns of spatial diversity in scientifically bred varieties, as indicated by the area planted to individual varieties. The concentration of varieties increases when a larger proportion of crop area is planted to fewer varieties. A considerable amount of evidence is available for wheat MVs. On the whole, this evidence suggests that across a wide range of countries, developing as well as industrialized, concentration in wheat varieties has not increased over time. Varietal concentration tends to be cyclical, often increasing soon after a particularly successful variety is introduced, but over the longer run the level of concentration usually decreases again (Smale, 1996). Although

Table 14.2. Latent Diversity in Gene Pools for Modern Cereal Varieties, as Measured by Coefficients of Parentage

Gene pool	Coefficient of parentage	Period or year	Source
Wheat			
USA hard red winter	0.30	1919	Cox, Murphy, and Rodgers (1986)
USA hard red winter	0.78	1984	Cox, Murphy, and Rodgers (1986)
USA soft red winter	0.92	1919	Cox, Murphy, and Rodgers (1986)
USA soft red winter	0.85	1984	Cox, Murphy, and Rodgers (1986)
USA hard red spring	0.74	Before 1940	van Beuningen (1993)
USA hard red spring	0.77	1941–1970	van Beuningen (1993)
USA hard red spring	0.78	1971–1990	van Beuningen (1993)
Canada hard red spring	0.81	Before 1940	van Beuningen (1993)
Canada hard red spring	0.67	1941–1970	van Beuningen (1993)
Canada hard red spring	0.42	1971–1990	van Beuningen (1993)
Australia state-by-state	0.60–0.81	1973	Brennan and Fox (1995)
Australia state-by-state	0.71–0.85	1993	Brennan and Fox (1995)
All Australia	0.81	1973	Brennan and Fox (1995)
All Australia	0.86	1993	Brennan and Fox (1995)
Europe ("elite" winter)	0.71		Plaschke, Ganal, and Röder (1995)
Yaqui Valley, Mexico	0.67	1972–1991	Souza et al. (1994)
Pakistan Punjab	0.74	1978–1990	Souza et al. (1994)
All India	0.90–0.95	1910–1920	Smale (1996)
All India	0.80–0.85	1980–1990	Smale (1996)
Rice			
USA long grain	0.79		Dilday (1990)
USA medium grain	0.70		Dilday (1990)
Philippines	0.86	1960–1994	de Leon (1994)
Eastern China	0.78–0.98		Widawsky (1996)
Latin America/Caribbean	0.80		Cuevas-Pérez et al. (1992)
Maize			
USA (elite Corn Belt inbreds)	0.86		Smith et al. (1990)
Of above, 1st heterotic group	0.73		Smith et al. (1990)
Of above, 2nd heterotic group	0.75		Smith et al. (1990)
USA (Corn Belt inbreds)	0.79		Bernardo (1993)
Europe, 18 flint inbreds	0.87		Messmer et al. (1993)
Europe, 11 dent inbreds	0.76		Messmer et al. (1993)

less evidence is available for other crops, studies of rice in Indonesia (Pardey *et al.*, 1992) and in eastern China (Widawsky, 1996) similarly fail to support the hypothesis that concentration in rice varieties is increasing. Data from the United States suggest that concentration also appears to have decreased in maize, wheat, soybeans, and cotton (Duvick, 1984).[4]

The fact that varietal concentration is declining over time does not mean that genetic diversity is increasing, however, since many "different" varieties grown by farmers may be closely related in the genetic sense. Information on varietal concentration can be combined with information on latent diversity through area-weighted diversity indices. Studies of wheat (Souza *et al.*, 1994; Smale, 1996) and rice (Widawsky, 1996) indicate that weighted diversity indices calculated for any given area are almost always lower than unweighted indices. Given that the evidence on long-term trends in the latent diversity of major modern varieties of cereal crops is basically inconclusive, it is perhaps not surprising that there is also little evidence of trends in artea-weighted latent diversity. Most studies conducted to date have covered limited time periods, however, and it is conceivable that future studies covering longer periods may show changes in area-weighted latent diversity.

14.4. DEMAND FOR SEED OF MODERN VARIETIES

14.4.1. Farmers' Seed Management Practices in Modern Varieties
Before examining the factors that influence demand for seed of MVs, it is useful briefly to consider farmers' seed management practices. Analysis of what is commonly referred to as "MV adoption" is often muddled by a failure to distinguish between the different reasons that motivate farmers to seek out and plant seed of MVs. We will distinguish between three different activities of farmers: MV adoption, MV replacement, and seed replacement. The purpose of distinguishing between MV adoption, MV replacement, and seed replacement is to help clarify how demand for MVs is expressed in the marketplace. As we shall see, certain attributes of seed affect the behavior of farmers in ways that produce highly variable levels of effective demand.

14.4.1.1. Modern Variety Adoption
A modern variety is "adopted" when a farmer who previously has grown only local varieties plants a modern variety for the first time. The adoption of MVs is usually motivated by the belief that planting the MV will bring greater benefits than those which would have been realized by continued use of the farmer's traditional variety. Adoption is thus motivated by the desire to gain access to the genetic material contained in seed. In most cases, MV adoption is associated with a substantial one-time yield increase representing the difference in genetic potential between a local variety (which may be a landrace) and a modern one. The much-publicized yield gains associated with the green revolutions in rice and wheat resulted from farmers' initial adoption of so-called "first-generation" MVs; in production environments favoring the production of rice and wheat, the initial yield gains frequently averaged as much as 35–40%.

14.4.1.2. Modern Variety Replacement

A modern variety is "replaced" when a farmer who already grows MVs stops planting one MV and starts planting another. Like MV adoption, MV replacement is motivated by the belief that switching to the new MV will bring greater benefits than those which would have been realized through continued use of the old MV. Replacement is thus also motivated primarily by the desire to gain access to the genetic material contained in the seed. Since the variety being replaced was itself developed by a formal plant breeding program, in most cases MV replacement is associated with a fairly modest yield increase, perhaps on the order of 5–10%. However, since MVs may be replaced regularly, at relatively frequent intervals, the cumulative effect of these changes can be quite large. Byerlee (1996) estimates that the total genetic gains attributable to regular replacement of the first-generation MVs of wheat in the Indian Punjab by second-, third-, and subsequent-generation MVs have averaged about 1% per year. As a result, the sum of incremental gains achieved during the post-green revolution period since 1975 has exceeded the gain associated with the initial adoption of MVs.

14.4.1.3. Seed Replacement

Seed is "replaced" whenever a farmer acquires new seed from an external source (as opposed to saving part of his or her own harvest for planting the following season). The adoption and replacement of MVs are always associated with seed replacement, because it is impossible to change varieties without changing seed. The converse is not true, however, in the sense that it is possible to replace seed without changing varieties. For MVs, this typically happens when farmers are satisfied with the performance of the variety and see no need to change their choice of germplasm, but for one reason or another do not wish to recycle seed. Replacement of MV seed is particularly common when farmers are growing hybrids, since hybrid seed cannot be recycled without the risk that its performance will decline. Farmers also replace seed of varieties, which may deteriorate in quality through repeated recycling to the point that replacement becomes desirable. Unlike MV adoption and replacement, seed replacement has nothing to do with gaining access to new genetic material. Lack of replacement may also contribute to greater intra-varietal diversity, especially in open-pollinated crops, as varieties become "rusticated" or farmers adapt them to their own conditions.

14.4.2. Factors Affecting Demand for Seed of Modern Varieties

What determines the demand for seed of MVs? Whether we are dealing with MV adoption, MV replacement, or simply seed replacement, the decision to use seed of MVs is generally based on the comparison of costs and benefits. Some of these costs and benefits are associated with factors that are valued explicitly by the market; these are invoked in conventional economic models of farmer decisionmaking. Other costs and benefits may be valued implicitly. Either type of costs and benefits can be treated within a general framework of utility-maximizing behavior.

Several levels of analysis are relevant to the demand for seed of MVs. Many studies of varietal use or MV adoption are based on sample surveys of a cross-section of farmers (including studies in part III of this volume). A smaller number of studies follow the example of Griliches' (1957) paper in attempting to understand and compare diffusion paths over time and across regions. Our classification of factors affecting the use of MV seed attempts to reconcile these two levels of analysis.

Factors that affect the demand for seed of MVs can be grouped loosely into two categories: variety-related factors and farmer characteristics. Variety-related factors include a mixture of economic and technical factors that determine the profitability of individual varieties and their suitability for each farmer's individual circumstances. Economic factors refer to any attribute or good over which a farmer's utility function is defined, whether or not it has an observable market price. Economic factors include the yield advantage associated with improved seed, the price of seed relative to the price of grain, and the cost of capital and learning. Although the importance of profitability is most evident in commercial production systems, profitability considerations also influence the decisions of subsistence-oriented farmers. Because price does not vary across the unit of analysis, however, cross-sectional studies of the adoption of varieties (as well as some econometric studies of seed supply and demand) often fail to provide empirical support to the hypothesis that seed price affects demand for MVs. When sufficient variability is present in the data, however, the influence of price on seed demand becomes evident (Heisey *et al.*, 1998).

Additional variety-related factors influencing the demand for seed of MVs can often be observed in detailed cross-sectional studies of MV use and adoption. They include the suitability of the MV for local production conditions, such as soil type (Bellon and Taylor, 1993) or suitability for local cropping systems and rotations (Byerlee, Akhtar, and Hobbs, 1987); the degree to which the MV satisfies local consumption preferences (Smale *et al.*, 1991; Bellon, Pham, and Jackson, 1997); the MV's ability to generate "joint" products such as grain and fodder (Traxler and Byerlee, 1993); and the perceived riskiness of the MV (Nowshirvani, 1971; Just and Zilberman, 1983). In addition to influencing farmers' decision to adopt MVs, the characteristics of varieties can explain why farmers often plant more than one variety of the same crop. Farmers who adopt MVs frequently also continue to plant traditional varieties. Smale, Just, and Leathers (1994) and Meng (1997) have reviewed and tested microeconomic explanations of this behavior.

Farmer characteristics that influence the demand for seed are often related to variation in the amount of human and/or financial capital present within households. These include such factors as the ability to acquire, process, and act upon information; the ability to mobilize financial resources for investment; and the ability (or willingness) to assume financial risk (Feder, Just, and Zilberman, 1985; Feder and Umali, 1993).

14.4.3. Implications for Genetic Diversity
In agricultural systems that rely on formal breeding programs for genetic material, which of the demand-side factors described above are most influential in determining the level of crop genetic diversity? In commercial cropping systems, farmers' choice of variety tends to be driven primarily by yields and prices, implying that these factors are most important in determining the level of crop genetic diversity found in farmers' fields. In the short run, yield is usually the primary consideration, which is why commercial farmers have become adept at choosing the highest yielding varieties

(Griliches, 1957; Byerlee, 1996). Constant gains in yield potential from research thus are likely to promote increased temporal diversity as farmers repeatedly replace older MVs with newer, higher-yielding ones (Heisey and Brennan, 1991). If increased spatial diversity is also to be promoted, the research system must not only produce a steady stream of ever-higher-yielding MVs, but each new generation of MVs must include a selection of varieties based on different genetic backgrounds. This procedure is contrary to what plant breeders understand, from experience, as the quickest way to achieve yield increases (Rasmusson, 1996).

In some circumstances, even in commercial farming areas, factors other than yield may be important in influencing farmers' choice of variety. For example, the Canadian Prairie wheat industry depends on the production of high-quality wheat that can be grown in relatively harsh environments. This is the reason genetic diversity in Canada is so low; Canadian wheat breeders usually incorporate new material through elaborate back-crossing programs that allow them to preserve the desired plant type and grain characteristics (van Beuningen, 1993; Thomas, 1996). Quality considerations may be particularly important in less commercialized areas where a substantial proportion of the farmer's crop is destined for consumption in the home.

14.5. SUPPLY OF SEED FOR MODERN VARIETIES

When a farmer decides to use seed of a modern variety, in order for that decision to be translated into action, the seed must be available when and where it is needed—and, in the case of purchased seed, the price must be remunerative. The dynamics of MV use thus depend not only on demand-side factors, but also on supply-side factors that determine seed availability and affordability.

14.5.1. Important Attributes of Seed
What factors influence the supply of seed of MVs? As with any good or service, the economic incentives to supply seed depend in large part on its characteristics. Unlike most other products, seed is really two things in one. First, it is a consumable input, i.e., something that can be combined with other inputs (e.g., land, labor, water, fertilizer, pesticide) to produce a crop. At the same time, maize seed is also a source of germplasm, i.e., a store of genetically encoded information that determines how other inputs combine and become transformed into useful products. Practically speaking, these two things are difficult to separate, in the sense that a farmer who purchases a bag of seed to acquire the consumable input needed for planting cannot avoid purchasing the genetically encoded information that is contained in the seed. Although seed considered as a consumable input and seed considered as a source of germplasm can have many attributes in common, they differ with respect to two attributes that turn out to be important in shaping production and consumption incentives: subtractability and excludability.

"Subtractability" refers to the degree to which use of a good or service by one person precludes use by another person. Seed considered as a consumable input has high subtractability, in the sense that two farmers cannot plant the same bag of seed. But seed considered as a source of germplasm has low subtractability, in the sense that if one farmer uses seed of a particular OPV or hybrid, this does not prevent other farmers from using the same OPV or hybrid.

"Excludability" refers to the ease with which the seller of a good or service can deny access to non-authorized users. Seed considered as a consumable input has high excludability, since sellers can deny access to non-authorized users simply by refusing to give them seed unless they pay the asking price. In contrast, the degree of excludability is different for seed considered as a source of germplasm. The germplasm in OPV seed has low excludability, because OPV is seed easily reproducible. But the germplasm in hybrid seed has high excludability, because hybrid seed can be reproduced only by those who know the pedigree or have access to the parent lines.

Considered as a source of germplasm, seed's low subtractability, coupled with variable excludability (low for OPVs, high for hybrids), suggest that the institutional arrangements needed effectively to develop, produce, and distribute improved seed will tend to vary. In the case of OPVs, incentives to produce and sell seed are undermined by the difficulty of appropriating profits from sales. In the case of hybrids, incentives to produce and sell seed are stronger, at least to the extent that the germplasm content can be kept out of the hands of potential competitors.

In summary, since the characteristics of seed can vary, the economic incentives to supply seed can vary as well. Incentives to supply seed are generally very low for non-hybrid crops, especially self-pollinating crops such as wheat and rice, because most farmers do not purchase commercial seed of these crops on a regular basis. If farmers cannot obtain seed from a relative or neighbor, they may initially purchase seed of a new variety, but after the initial cropping season it is usually economical to recycle seed from their own harvest for planting in the next season.[5] Economic incentives to supply seed are usually much greater for hybrid crops, because farmers tend to purchase commercial seed of these crops quite frequently. As a result, private seed companies tend to be very active in producing and selling seed of hybrids, but they are far less active when it comes to selling non-hybrid seed.

14.5.2. Economics of Seed Production

Because of the high level of private-sector participation in hybrid seed markets, most empirical studies of seed industry organization and performance have focused on industries involving hybrid crops. Many of the factors that influence the supply of hybrid seed were first identified by Griliches (1957) in his pioneering study of the diffusion of hybrid maize in the United States. Based on statistical analysis, Griliches determined that the diffusion of hybrid maize throughout the Unites States Corn Belt was driven by profitability considerations related to the structure of the seed market. Griliches' concepts were later adapted and modified by Heisey et al. (1998) in analyzing hybrid seed adoption patterns in a sample of 32 developing countries. Although these studies focused on the adoption and diffusion of hybrid maize seed, the determinants of improved seed supply are similar for other crops.

Building on Griliches' earlier work, Heisey et al. (1998) argued that for most private seed companies the market entry decision depends on five considerations: (1) seed production costs, (2) market structural conditions, (3) the organization of the seed industry, (4) the cost of research innovation, and (5) the political importance of the crop.

14.5.2.1. Seed Production Costs

Seed production costs are determined by technical and economic factors. Technical factors affect seed production costs by imposing physical limits on seed yields, irrespective of the management ability of the seed grower. For example, seed yields of the parental materials used to produce different types of hybrids vary considerably. Seed production costs are also influenced by economic factors, especially the prices of key inputs such as labor, land, and capital.

14.5.2.2. Market Structural Conditions

The bigger the potential market for commercial seed, the larger the potential profits for seed organizations—and the greater the incentives for them to invest in seed research, seed production, and seed distribution activities. To the extent that these activities are characterized by economies of scale, per-unit production costs will decline as market size increases. The composition of the market also influences supply incentives, since distribution costs per unit of seed sold are related to the average crop area per farm. Every seed sale involves some fixed transactions costs, so the fixed cost component per unit of seed decreases as the size of the sale increases. Thus, other things being equal, a market made up of large-scale producers will be more attractive to seed companies than one made up of small-scale producers.

14.5.2.3. Organization of the Seed Industry

In theory, seed industries characterized by large numbers of small companies actively competing for market share should be more competitive than industries made up of small numbers of large companies capable of exercising monopoly or oligopoly power; this competition should be favorable for farmers. In practice, the relationship between seed industry concentration and competitiveness is far from deterministic. It is instructive to consider the experience of the United States, where maize seed industries have become concentrated in the hands of a few companies, but where the battle for market share is intense. It may actually be desirable to have a seed industry made up of fewer, larger companies if these companies can capture economies of scale and pass them along to farmers via lower seed prices.

14.5.2.4. Cost of Research Innovation

The cost of research innovation can be expected to change through time, although the rate and even the direction of change are not easy to predict. Technological advances may reduce the cost of innovation, as for example when the basic principles of hybridization were discovered. Similarly, the cost of innovation for a particular region or group of farmers may decline when there are opportunities to capture spillover benefits from research done elsewhere. But other factors may work to increase the cost of innovation. For some crops or regions, most of the "easy" gains from conventional plant breeding research may have already been realized. In the United States, for example, investment in maize breeding research has been increasing exponentially, while the genetic potential of commercial maize hybrids has been increasing only linearly (Byerlee and López-Pereira, 1994).

14.5.2.5. Importance of the Crop
Another factor that can influence the supply of commercial seed is the economic or political importance of the crop, since governments are more likely to enact measures designed to encourage domestic production when a crop is economically or politically important. In developing countries, this typically occurs when a crop is regarded as a major food staple, an important commercial crop, or a strategic export crop.

14.5.3. Dynamics of Commercial Seed Industries
For a given crop in a given country, the five factors described in section 14.5.2 influence the supply of commercial seed by determining the economic potential of the overall market. If the combined effect of the factors is to create an attractive potential market, profit-oriented firms will be encouraged to enter the market. Once the market entry decision has been made, each company must develop a commercial strategy that will allow it to compete successfully in the marketplace. For private companies, the commercial strategy usually will involve the following activities: identifying target environments, developing superior products (varieties or hybrids), producing commercial seed, convincing farmers to buy the seed, and delivering the seed efficiently at an affordable price.

Initially at least, it is unlikely that a single organization will pursue all of these activities. The locus of seed research, seed production, and seed distribution activities frequently varies by type of crop, by the structure of demand, and by the degree of maturity of the seed industry. Certain activities (e.g., basic research, farmer education, production of non-hybrid seed) offer limited profit opportunities and therefore tend to attract little interest from private firms; responsibility for these activities tends to be assumed by public-sector organizations that are able to operate only with the help of government subsidies. Other activities (e.g., applied research, hybrid seed production) offer significant profit opportunities and attract considerable attention from private firms; these activities tend to be characterized by an absence of public-sector participation.

The roles played by different types of organizations (e.g., public, private, participatory) change as the seed industry evolves and matures. A series of recent case studies focusing on the emergence of national maize seed industries in Africa, Asia, and Latin America suggest that seed industry development often follows the same general path (Morris, 1998). The findings of these case studies lend support to lifecycle theories of industrial development advanced by a number of seed industry analysts (Douglas, 1980; Desai, 1985; Pray and Ramaswami, 1991; Rusike, 1995; Dowswell, Paliwal, and Cantrell, 1996). Four stages of growth are commonly described, including pre-emergence, emergence, expansion, and maturity. In reality these "stages" represent arbitrarily selected points along a growth continuum (Table 14.3).

14.6. IMPLICATIONS FOR CROP DIVERSITY

The lifecycle view of seed industry development provides a number of important insights about crop diversity. First, where conditions encourage a high level of private-sector participation in the seed industry, the seed industry will tend to become increasingly concentrated and vertically integrated, as profit-oriented firms seek to capture economic efficiencies. If a small number of firms come to dominate a particular seed industry, the choice of MVs offered to farmers may decline. A reduction in the number of available MVs will tend to reduce the level of inter-varietal diversity found at the farm level.

Second, where conditions do not encourage a high level of private-sector participation in the seed industry, public organizations will be forced to assume a leading role in research, seed production, and seed distribution. In such cases, the choice of MVs offered to farmers will depend on the following factors:

1. The level of public investment in plant breeding research (and the productivity of plant breeders) will determine the number of new varieties that are developed.
2. Official varietal testing and release policies will determine what proportion of these varieties are approved for commercial use.

Table 14.3. Characteristics Associated with the Stages of Maize Seed Industry Development

	Stage 1: Pre-industrial	Stage 2: Emergence	Stage 3: Expansion	Stage 4: Maturity
Orientation of agriculture	Subsistence	Semi-subsistence	Mostly commercial	Completely commercial
Predominant seed technology	OPVs	OPVs, some hybrids	Some OPVs, hybrids	Hybrids
Seed procurement practices	On-farm production, farmer-to-farmer exchange	On-farm production, farmer-to-farmer exchange,some purchasing	Frequent purchasing	Annual purchasing
Seed production	On-farm	On-farm, public organizations	On farm, public organizations, private companies (national)	Private companies (global)
Seed market coverage	Local	Local, regional	Local, regional, national	Local, regional, national, global
Sources of seed information	Direct experience, other farmers	Public organizations	Private seed companies	Private seed companies
Locus of seed research and development	On farm	Public organizations	Public and private organizations	Public and private organizations (specialized)
Supporting legal systems	Customary law	Civil	Commercial (domestic)	Commercial (global)
Intellectual property rights	None	None	Trade secrets	Plant variety protection, patents

Source: Morris, Rusike, and Smale (1998).

3. The technical proficiency and managerial skills of state seed production agencies will determine whether high-quality seed of the approved varieties is made available to farmers when and where it is needed.
4. Input pricing policies will determine whether or not this seed is offered at remunerative prices.

Assuming that the number of available MVs (and the quantities available of high-quality seed) affect MV adoption rates and diffusion patterns, the performance of public breeding programs and public seed production agencies will tend to influence the level of inter-varietal diversity at the farm level.

Third, because it is costly to develop and promote new varieties, profit-oriented seed companies have incentives to encourage as many farmers as possible to plant the same variety for the longest possible time. To the extent that particular varieties succeed in dominating markets, spatial and temporal diversity will tend to be reduced.

Fourth, to be successful, seed companies must convince farmers to buy their products. Seed companies use a number of strategies to convince farmers that their products are superior to those of their competitors. In the case of publicly bred varieties, which can be sold by many companies, the germplasm content is not a distinguishing feature, so companies promote other characteristics of the seed (e.g., cleanliness, viability, quality of packaging materials, price). In the case of proprietary materials, the germplasm content is often the distinguishing feature, so companies tend to promote the performance of the germplasm (e.g., yield, grain quality, maturity, insect or disease resistance, drought tolerance). Since farmers cannot easily determine the germplasm content of seed, however, companies may not necessarily offer products that really are different from those offered by their competitors. If rival seed companies promote products that are genetically similar, farm-level genetic diversity may be quite narrow even when farmers are growing a wide range of varieties.

14.7. SUMMARY AND DISCUSSION

This chapter has examined crop diversity in rice, wheat, and maize, focusing on spatial, temporal, and genealogical dimensions of MV adoption and diffusion in commercial production systems. Based on admittedly imperfect measures, it has not been possible to determine any clear trends in the level of genetic diversity found within sets of related MVs (e.g., all of the semidwarf wheat varieties released within a given country). Until better measures of crop genetic diversity are developed, it will be difficult to establish whether levels of diversity in farmers' fields are increasing, remaining constant, or decreasing. Similarly, until ways are found to place an economic value on genetic diversity, it will be hard to say whether existing levels of crop diversity are "too high," "about right," or "too low."

Factors affecting the demand for and supply of seed were examined in an attempt to explain what drives the MV adoption and diffusion process. Knowledge of these factors allows us to anticipate future trends in MV use, which will influence the level of diversity present in farmers' field. Knowledge of the factors that drive the MV adoption and diffusion process also allows us to formulate strategies to avert a socially undesirable narrowing of crop diversity through selected policy interventions designed to modify the behavior of key actors.

As we have seen, in commercial cropping systems demand for seed of MVs is driven mainly by factors that affect the expected profitability of seed use. Whenever large numbers of farmers possess similar resource endowments, knowledge, and technical skills, and assuming that they face similar incentives, many will be influenced to select the same MVs. To the extent that concentration of area among varieties is associated with reduced genetic diversity, this could lead to a reduction in overall crop diversity measured across the entire population of farmers.

This situation—in which individuals acting in their own interests collectively produce a result that is undesirable for the group as a whole—can be characterized as a "social trap" (Schmid, 1987; Cornes and Sandler, 1986). Social traps arise when there is an incongruity between individual incentives and group outcomes. In the current context, if all farmers within a specified area select the same MV in the hope of maximizing production on their own farms, the risk of a catastrophic disease epidemic increases for everybody (Heisey et al. 1997). Social traps tend to be difficult to avoid; to maximize the collective welfare of the group, individuals must be induced to act in ways that run counter to their own immediate self-interest. Even if they understand the link between individual actions and collective outcomes, individuals will still have strong incentives to "cheat" for their own benefit.

The way to avoid falling into a social trap is to get individuals to modify their behavior through education, economic inducement, or coercion. From a policy perspective, which if any of these alternatives is likely to be effective in convincing farmers to act in ways that enhance genetic diversity in the crops they grow?

Education can certainly help to sensitize farmers about the importance of crop genetic diversity, but profit-motivated farmers will in most instances be unwilling to plant diversity-enhancing varieties to benefit society as a whole if the result will be a reduction in their own benefits. Thus education per se will usually be ineffective.

Economic inducements might be more effective. For example, farmers who agree to plant diversity-enhancing varieties could receive financial compensation for the associated production loss. Unless production is concentrated in the hands of a small number of large-scale producers, however, the administrative costs of implementing a program of economic compensation are likely to be very high, ruling it out as a feasible policy option.

Coercion has been tried in a number of countries. China before the reform of the 1980s is a classic example of a country where production decisions were centrally administered and farmers were ordered what crops (and which varieties) to plant. The ineffectiveness of the system is legendary. Most households invested the minimum effort needed to satisfy mandatory production targets and diverted their excess resources into private plots; at the same time, the system was costly in the sense that government was forced to invest considerable resources into administration and enforcement measures. Less authoritarian systems have produced similar results. For example, the government of Pakistan annually publishes a list of wheat varieties that farmers are prohibited from growing because the varieties are vulnerable to disease. In general, the bans have proven impossible to enforce, since the cost of identifying and sanctioning individual transgressors is prohibitive.

Demand-side policy measures in and of themselves thus appear to offer few workable opportunities for influencing the aggregate level of cultivar diversity found in farmers' fields. So what about the supply side? Our review of the factors that shape the incentives to supply seed of MVs led us to conclude that profit-motivated companies have incentives to promote the smallest possible number of MVs to the largest possible number of customers for the longest possible time. Under these circumstances, utility-maximizing behavior on the part of individual seed companies is likely to lead to decreasing spatial, temporal, and (perhaps) latent diversity, as measured by pedigrees.

What are the prospects of influencing the supply side of the seed demand–supply equation to encourage cultivar diversification and/or enhance crop genetic diversity? The goal in this case is to shift a negative externality (vulnerability to crop losses resulting from insufficient genetic diversity) from the farmers who adopt the MVs back to the seed suppliers themselves. As in the case of the demand-side social trap, the same three basic options are available for modifying the behavior of seed suppliers.

Education can help to sensitize plant breeders and seed producers to the dangers of reduced genetic diversity. Historically, one of the main factors preserving current levels of genetic diversity in cereals has been plant breeders' continual efforts to seek out and exploit new sources of desirable traits. Whether or not additional knowledge would induce breeders to increase the supply of genetically diverse MVs will depend on the cost of breeding greater genetic diversity into commercial varieties. If a significant cost is associated with breeding greater diversity into plants, then researchers and seed companies are unlikely to bear that cost voluntarily, since they would receive little direct benefit. On the other hand, if plant breeders have the ability to choose costlessly among a wide range of genetically diverse germplasm sources, then they might be willing to include genetic diversity as an explicit criterion in their selection indices. Experience suggests that the source materials available to breeders are not equal in terms of performance, however, so it is likely that breeding greater diversity into commercial varieties will to come at some cost to other performance characteristics.

If seed suppliers cannot be induced voluntarily to increase the level of genetic diversity in commercial varieties, can they be induced to do so through economic incentives or coercion? In principle, either strategy could work. Plant breeders presumably could be paid to develop genetically diverse MVs, and seed companies could be paid to produce seed of these MVs. Alternatively, plant breeders could be required by law to develop genetically diverse MVs, and seed companies could be mandated to produce seed of these MVs. Unfortunately, both strategies have a major drawback. Just because genetically diverse MVs are available does not mean farmers will adopt them. If enhanced genetic diversity is associated with a significant cost in performance, farmers are unlikely to use genetically diverse MVs voluntarily. Thus we are back to the demand-side problem discussed earlier.

In summary, factors affecting the demand and supply of improved seed suggest that as agriculture becomes commercialized, greater areas could be planted to a relatively small number of MVs with slower turnover. To date, any trend in this direction has apparently been offset by the increasing pedigree complexity of recently released MVs, made possible because scientific advances have enabled plant breeders to exploit a wider range of source materials and to target MVs more precisely to specific agroecological niches. Should future MV adoption and diffusion processes result in a socially undesirable level of crop diversity, however, it may become desirable to introduce policy measures designed to encourage diversity. Although economic incentives and/or mandatory regulations may offer workable strategies for getting seed suppliers to increase the range of genetically diverse MVs on offer, inducing farmers to plant specific MVs for the express purpose of increasing social welfare will be difficult when it is not in their own interest to do so.

Notes

1 The only exception is in the case of first-generation hybrid maize grown from purchased seed, which is genetically uniform.
2 The difficulty of defining precisely what constitutes an "improved" OPV is further complicated by another factor: given the propensity of maize plants to cross-pollinate, to what extent have local varieties and landraces been "improved" as the result of exposure to scientifically bred cultivars?
3 Although not shown in Table 14.1, in developing countries the area under MVs is generally smaller for other cereals. However, there are some significant exceptions. For example, in India just over half of the sorghum area and just under half of the millet area are planted to improved OPVs and hybrids (Fertiliser Association of India, 1995).
4 Duvick's tentative findings regarding declining cultivar concentration in United States wheat are given stronger support over a longer period by Dalrymple (1988).
5 A number of European countries have recently introduced measures to overcome this problem by requiring farmers to pay an annual licensing fee for the use of farmer-produced varietal seed.

References

Almekinders, C. J. M., N. P. Louwaars, and G. H. de Bruijn. 1994. Local seed systems and their importance for an improved seed supply in developing countries. *Euphytica* 78: 207–216.
Bellon, M. R., J. L. Pham, and M. T. Jackson. 1997. Genetic conservation: A role for rice farmers. In N. Maxted, B.V. Ford-Lloyd and J.G. Hawkes (eds.), *Plant Conservation: The In Situ Approach*. London: Chapman and Hall.
Bellon, M. R., and J. E. Taylor. 1993. Folk soil taxonomy and partial adoption of new seed varieties. *Economic Development and Cultural Change* 41: 763–786.
Bernardo, R. 1993. Estimation of coefficient of coancestry using molecular markers in maize. *Theoretical and Applied Genetics* 85: 1055–1062.
Brennan, J. P., and D. Byerlee. 1991. The rate of crop varietal replacement on farms: Measures and empirical results for wheat. *Plant Varieties and Seeds* 4: 99–106.
Brennan, J. P., and P. N. Fox. 1995. *Impact of CIMMYT Wheats in Australia: Evidence of International Research Spillovers*. Economics Research Report No. 1/95. Wagga Wagga: NSW Agriculture.
Byerlee, D. 1996. Modern varieties, productivity, and sustainability: Recent experience and emerging challenges. *World Development* 24: 697–718.
Byerlee, D., M. R. Akhtar, and P. R. Hobbs. 1987. Reconciling conflicts in sequential cropping patterns through plant breeding: The example of cotton and wheat in Pakistan's Punjab. *Agricultural Systems* 24: 291–304.
Byerlee, D., and M. A. López-Pereira. 1994. *Technical Change in Maize Production: A Global Perspective*. CIMMYT Economics Working Paper 94-02. Mexico, D.F.: International Maize and Wheat Improvement Center (CIMMYT).
Byerlee, D., and P. Moya. 1993. *Impacts of International Wheat Breeding Research in the Developing World, 1966–1990*. Mexico, D.F.: International Maize and Wheat Improvement Center (CIMMYT).
CIMMYT (International Maize and Wheat Improvement Center). 1993. *1992/93 CIMMYT World Wheat Facts and Trends. The Wheat Breeding Industry in Developing Countries: An Analysis of Investments and Impacts*. Singapore: CIMMYT.
CIMMYT (International Maize and Wheat Improvement Center). 1994. *1993/94 CIMMYT World Maize Facts and Trends. Maize Seed Industries, Revisited: Emerging Roles of the Public and Private Sectors*. Mexico, D.F.: CIMMYT.

CIMMYT (International Maize and Wheat Improvement Center). 1996. *CIMMYT World Wheat Facts and Trends 1995/96: Understanding Global Trends in the Use of Wheat Diversity and International Flows of Wheat Genetic Resources*. Mexico, D.F.: CIMMYT.

Cornes, R., and T. Sandler. 1986. *The Theory of Externalities, Public Goods, and Club Goods*. Cambridge: Cambridge University Press.

Cox, T. S., J. P. Murphy, and D. M. Rodgers. 1986. Changes in genetic diversity in red winter wheat regions of the United States. *Proceedings of the National Academy of Sciences (U.S.A.)* 83: 5583–5586.

Cuevas-Pérez, F. E., E. P. Guimareas, L. E. Berrio, and D. I. González. 1992. Genetic base of irrigated rice in Latin America and the Caribbean. *Crop Science* 32: 1054–1059.

Dalrymple, D. G. 1988. Changes in wheat varieties and yields in the United States, 1919–1984. *Agricultural History* 62: 20–36.

de Leon, J. C. 1994. Genetic relationships among Philippine-bred rice varieties as determined by pedigree- and morphology-based measures. Ph.D. thesis, University of the Philippines, Los Baños, Philippines.

Desai, G. 1985. Understanding marketing environment for seed industry. Teaching note. Ahmedabad: Centre for Management in Agriculture, Indian Institute of Management.

Dilday, R. H. 1990. Contribution of ancestral lines in the development of new cultivars of rice. *Crop Science* 30: 905–911.

Douglas, J. E. (ed.). 1980. *Successful Seed Programs: A Planning and Management Guide*. Boulder: Westview Press.

Dowswell, C. R., R. L. Paliwal, and R. P. Cantrell. 1996. *Maize in the Third World*. Boulder: Westview Press.

Duvick, D. N. 1984. Genetic diversity in major farm crops on the farm and in reserve. *Economic Botany* 38: 161–178.

Duvick, D. N. 1992. Genetic contributions to advances in yield of U.S. maize. *Maydica* 37: 69–79.

Duvick, D. N. 1998. United States. In M. L. Morris (ed.), *Maize Seed Industries in Developing Countries*. Boulder: Lynne Rienner and International Maize and Wheat Improvement Center (CIMMYT).

Evenson, R. E., and D. Gollin. 1997. Genetic resources, international organizations, and improvement in rice varieties. *Economic Development and Cultural Change* 45: 471–500.

Feder, G., R. E. Just, and D. Zilberman. 1985. Adoption of agricultural innovations in developing countries: A survey. *Economic Development and Cultural Change* 33: 255–297.

Feder, G., and D. L. Umali. 1993. The adoption of agricultural innovations: A review. *Technological Forecasting and Social Change* 43: 215–239.

Fertiliser Association of India. 1995. *Fertiliser Statistics, 1994–95*. New Delhi: The Fertiliser Association of India.

Griliches, Z. 1957. Hybrid corn: An exploration in the economics of technical change. *Econometrica* 25: 501–522.

Heisey, P. W. 1990. *Accelerating the Transfer of Wheat Breeding Gains to Farmers: A Study of the Dynamics of Varietal Replacement in Pakistan*. CIMMYT Research Report No. 1. Mexico, D.F.: International Maize and Wheat Improvement Center (CIMMYT).

Heisey, P. W., and J. P. Brennan. 1991. An analytical model of farmers' demand for replacement seed. *American Journal of Agricultural Economics* 73: 1044–1052.

Heisey, P. W., D. Byerlee, M. L. Morris, and M. A. López-Pereira. 1998. Economics of adoption of hybrid maize. In M. L. Morris (ed.), *Maize Seed Industries in Developing Countries*. Boulder: Lynne Rienner and International Maize and Wheat Improvement Center (CIMMYT).

Heisey, P. W., M. Smale, D. Byerlee, and E. Souza. 1997. Wheat rusts and the costs of genetic diversity in the Punjab of Pakistan. *American Journal of Agricultural Economics* 79: 726–737.

IRRI (International Rice Research Institute). 1995. *World Rice Statistics, 1993–94*. Manila: IRRI.

Jatileksono, T., and K. Otsuka. 1993. Impact of modern rice technology on land prices: The case of Lampung in Indonesia. *American Journal of Agricultural Economics* 75: 652–665.

Just, R. E., and D. Zilberman. 1983. Stochastic structure, farm size and technology adoption in developing agriculture. *Oxford Economics Papers* 35: 307–328.

López-Pereira, M. A., and M. L. Morris. 1994. *Impacts of International Maize Breeding Research in the Developing World, 1966–90*. Mexico, D.F.: International Maize and Wheat Improvement Center (CIMMYT).

Meng, E. C. H. 1997. Land allocation decisions and *in situ* conservation of crop genetic resources: The case of wheat landraces in Turkey. Ph.D. thesis, University of California, Davis, California.

Messmer, M. M., A. E. Melchinger, R. G. Herrman, and J. Boppenmaier. 1993. Relationships among early European maize inbreds: II. Comparison of pedigree and RFLP data. *Crop Science* 33: 944–950.

Morris, M. L. (ed.). 1998. *Maize Seed Industries in Developing Countries*. Boulder: Lynne Rienner and International Maize and Wheat Improvement Center (CIMMYT).

Morris, M. L., J. Rusike, and M. Smale. 1998. Maize seed industries: A conceptual framework. In M. L. Morris (ed.), *Maize Seed Industries in Developing Countries*. Boulder: Lynne Rienner and International Maize and Wheat Improvement Center (CIMMYT).

Nowshirvani, V. F. 1971. Land allocation under uncertainty in subsistence agriculture. *Oxford Economics Papers* 23: 445–455.

Pardey, P. G., J. M. Alston, J. E. Christian, and S. Fan. 1996. *Summary of A Productive Partnership: The Benefits from U.S. Participation in the CGIAR*. EPTD Discussion Paper No. 18. Washington, DC: International Food Policy Research Institute (IFPRI) and University of California, Davis.

Pardey, P. G., R. K. Lindner, E. Abdurachman, S. Wood, S. Fan, W. M. Eveleens, B. Zhang, and J. M. Alston. 1992. *The Economic Returns to Indonesian Rice and Soybean Research*. An AARD/ISNAR Report. The Hague: International Service for National Agricultural Research (ISNAR).

Plaschke, J., M. W. Ganal, and M. S. Röder. 1995. Detection of genetic diversity in closely related bread wheat using microsatellite markers. *Theoretical and Applied Genetics* 91: 1001–1007.

Pray, C. E., and B. Ramaswami. 1991. *A Framework for Seed Policy Analysis in Developing Countries*. Washington, DC: International Food Policy Research Institute (IFPRI).

Rasmusson, D. C. 1996. Germplasm is paramount. In M. P. Reynolds, S. Rajaram, and A. McNab (eds.), *Increasing Yield Potential in Wheat: Breaking the Barriers*. Mexico, D.F.: International Maize and Wheat Improvement Center (CIMMYT).

Rusike, J. 1995. An institutional analysis of the maize seed industry in southern Africa. Ph.D. thesis, Michigan State University, East Lansing, Michigan.

Schmid, A. A. 1987. *Property, Power, and Public Choice*. Second edition. New York: Praeger.

Smale, M. 1996. *Understanding Global Trends in the Use of Wheat Diversity and International Flows of Wheat Genetic Resources*. CIMMYT Economics Working Paper 96-02. Mexico, D.F.: International Maize and Wheat Improvement Center (CIMMYT).

Smale, M., R. E. Just, and H. D. Leathers. 1994. Land allocation in HYV adoption models: An investigation of alternative explanations. *American Journal of Agricultural Economics* 76: 535–546.

Smale, M., with Z. H. W. Kaunda, H. L. Makina, M. M. M. K. Mkandawire, M. N. S. Msowoya, D. J. E. K. Mwale, and P. W. Heisey. 1991. Chimanga cha Makolo, *Hybrids and Composites: An Analysis of Farmers' Adoption of Maize Technology Adoption in Malawi, 1989–1991*. CIMMYT Economics Working Paper 91-04. Mexico, D.F.: International Maize and Wheat Improvement Center (CIMMYT).

Smith, O. S., J. S. C. Smith, S. L. Bowen, R. A. Tenborg, and S. J. Wall. 1990. Similarities among a group of elite maize inbreds as measured by pedigree, F_1 grain yield, grain yield heterosis, and RFLPs. *Theoretical and Applied Genetics* 80: 833–840.

Souza, E., P. N. Fox, D. Byerlee, and B. Skovmand. 1994. Spring wheat diversity in irrigated areas of two developing countries. *Crop Science* 34: 7 74–783.

Sperling, L., U. Scheidegger, and R. Buruchara. 1996. *Designing Seed Systems with Small Farmers: Principles Derived from Bean Research in the Great Lakes Region of Africa*. Agricultural Research and Extension Network Paper No. 60. London: Overseas Development Agency.

Thomas, N. 1996. *Use of IARC Germplasm in Canadian Crop Breeding Programs: Spillovers to Canada from the CGIAR and Some Outflows from Canada*. Ottawa: Canadian International Development Agency (CIDA).

Traxler, G., and D. Byerlee. 1993. A joint-product analysis of the adoption of modern cereal varieties in developing countries. *American Journal of Agricultural Economics* 75: 981–989.

van Beuningen, L. T. 1993. Genetic diversity among North American spring wheat cultivars as determined from genealogy and morphology. Ph.D. thesis, University of Minnesota, St. Paul, Minnesota.

Widawsky, D. A. 1996. Rice yields, production variability, and the war against pests: An empirical investigation of pesticides, host-plant resistance, and varietal diversity in Eastern China. Ph.D. thesis, Stanford University, Palo Alto, California.

15 COLLABORATIVE PLANT BREEDING AS AN INCENTIVE FOR ON-FARM CONSERVATION OF GENETIC RESOURCES: ECONOMIC ISSUES FROM STUDIES IN MEXICO

M. Smale, D. Soleri, D. A. Cleveland, D. Louette,
E. Rice, J.-L. Blanco, and A. Aguirre

15.1. CONTEXT

15.1.1. Perspectives on Conservation

One characteristic of classical crop improvement and genetic resource conservation programs is their physical and temporal distance from one another as well as from the farmers who are their clients. Genetic resources in breeders' working collections or conserved *ex situ* in gene banks are used in crosses, and selection in segregating populations is carried out under experimental conditions. The resulting varieties or advanced lines are eventually tested in a range of sites, but often testing does not include the fields of farmers, especially small-scale farmers in environments beset by biotic and abiotic stresses. The finished products are released after years of research and intended for use over extensive geographical areas. Varieties are developed to be highly responsive to certain growing conditions. Controlling—and thus simplifying—growing environments with respect to water supply, soil fertility, pests, and diseases has been an effective, efficient strategy. Through this strategy, the utilization of genetic resources conserved *ex situ* expands from the local to the global arena, which potentially increases their economic value. Furthermore, the costs of conserving genetic resources of a given crop are not borne directly by the farmers who produce the crop. In this classical system, the goal of conservationists is to preserve maximum allelic diversity in crop populations that they define as having global importance. We refer to this as the "conservationist perspective."

In traditional, low-resource farming communities located in marginal, variable environments, the crop populations that endure are those that meet production and consumption standards and that possess the genetic variability to respond to continual changes in farmers' needs and growing environments. Farmers in such communities choose among crop populations and select within them to meet their needs, given their economic and environmental constraints. Since these needs are defined both in the present and the future, the crop populations maintained by these communities serve both production and "conservation" functions simultaneously, as these are locally defined. Often, genetic diversity eventually has a negative effect on productivity through reduced adaptation, so for farmers the optimal level of diversity may be less than it is for conservationists, even in the absence of other constraints. We refer to this as the "farmer perspective."

When these perspectives are compared, two points emerge that have particular relevance to economic incentives and conservation policy. First, in the locally based system compared to the more classical system, farmers themselves directly bear the costs of conservation. Second, the crop populations that farmers seek to maintain locally are not likely to be those that preserve maximum allelic diversity among populations identified as global conservation targets.

Soleri and Smith (1995) first identified this contrast in analyzing the prospects for conserving traditional crop varieties *in situ*. The same contrast recurs in the context of collaborative plant breeding. "Collaborative plant breeding" is a recently named, but not altogether new, approach to crop improvement for meeting the needs of agricultural communities. It differs technically from more classical crop improvement and conservation efforts, although it may serve similar goals.

The idea behind collaborative plant breeding is that the biological effectiveness of plant breeding and/or the breadth of its social impact can be enhanced by drawing farmers into developing varieties with professional plant breeders or, conversely, by bringing professional plant breeders closer to farmers' local conditions and selection and maintenance practices. The collaborative plant breeding idea has emerged at the same time as, and has been encouraged by, recent interest in conservation *in situ* (see chapters in part III). Potentially this sets the stage for a conflict between the farmer and conservationist perspectives. As a means to improve crops locally, collaborative plant breeding is intended to serve the goals of farmers; as a proposed incentive for *in situ* conservation, collaborative plant breeding serves the goal of conservationists. For example, suppose that farmers considered a collaborative plant breeding initiative to be less successful in terms of crop improvement than conservationists considered it to be in terms of maintaining genetic diversity. Effectively, farmers would be "paying," at a net loss to themselves, for global conservation.

15.1.2. The Goals of Collaborative Plant Breeding

Farmers are themselves plant breeders, but we use the term "professional plant breeder" to refer to persons who practice plant breeding full-time and are paid to do so. Farmers are compensated for their plant breeding efforts indirectly—through the enhanced value of their crop in money or in kind. The livelihood of many small-scale farmers depends not only on the efficacy of their choice and breeding of crop varieties in terms of food production on their own farms, but also on their engagement in activities other than farming. In comparison, the success of most professional plant breeders has traditionally depended on the development of varieties acceptable to a significant proportion of farmers in selected target areas.

Collaborative plant breeding refers to a range of crop improvement activities defined by the relative involvement of farmers and professional plant breeders in the development of new crop varieties. To date, most collaborative plant breeding efforts have attempted to bring farmers into an established breeding agenda earlier than in classical programs, where they enter the picture when choosing among released varieties supplied to them through the seed industry. Building on Biggs' (1989) typology of farmer participation in national agricultural research systems, Witcombe and Joshi (1996) have subdivided this approach to collaborative plant breeding into two categories. Participatory varietal choice operates at the inter-varietal level. In their own fields or in on-station demonstrations, farmers choose "finished" varieties from among those offered by plant breeding programs before release. Participatory plant breeding occurs at the intra-varietal level. In the earlier stages of cultivar development, farmers are included in the choice of plant characteristics to improve, crop populations, and breeding techniques. Within these two broad approaches, there is a wide range of potential farmer and professional involvement, depending on the setting and objectives of the research. Experience is limited with both categories of activities, however. An alternative approach is to ask breeders to contribute to an established, farmer-based breeding or selection regime. Here, although the range of potential contributions by both farmers and professionals is also wide, experience is even more limited.

Among the multiple goals of collaborative plant breeding that have been proposed in the literature, two are of primary interest in this chapter. The first is to enhance the contribution of farmers' varieties to their production goals, by improving yield, stability, or post-harvest traits related to storage, processing, or consumption quality. Through selecting for specific adaptation, breeders may better meet the needs of poor farmers in stressed environments who have not yet received benefits from improved varieties (Ceccarelli et al., 1997). Collaborative plant breeding can enhance the effectiveness of plant breeding programs by increasing the likelihood that selection criteria and methods are relevant for local environmental demands as well as farmers' needs (van Oosterom, Whitaker, and Weltzien, 1996; Weltzien, Whitaker, and Anders, 1996). Similarly, collaborative plant breeding could enable plant breeders to help farmers improve the efficiency of their own plant breeding.

The second proposed goal of collaborative plant breeding is to facilitate the conservation of crop genetic diversity in farmers' fields and storage. Proponents of this approach argue that while professional plant breeders have conventionally sought to develop fewer varieties adapted to a wider range of locations, participatory breeding can support the maintenance of more diverse, locally adapted plant populations (Berg, 1995; Cleveland, Soleri, and Smith, 1994; Witcombe and Joshi, 1995). Collaborative plant breeding can serve as a link between agricultural development and genetic resource conservation (Eyzaguirre and Iwanaga, 1996). Qualset *et al.* (1997) have proposed the improvement of farmers' breeding methods as an incentive for local conservation. For example, local varieties may provide the basis for breeding and introgression of genes from exotic sources.

The biological validity of this second proposed role has not yet been documented. Nor has an economic analysis yet been published on collaborative plant breeding initiatives, according to Witcombe (1997). One reason is that those who work closely with farm communities, such as small non-governmental organizations, are typically too busy implementing projects to document them. Another is that measuring impact with conventional cost-benefit techniques is not acceptable to many who are involved in such initiatives, and more comprehensive methods are needed.

15.1.3. Scope and Purpose of This Chapter

For economists, a fundamental issue surrounding collaborative plant breeding efforts concerns farmers' incentives to participate in those efforts. In this chapter, we explore farmers' incentives to engage in collaborative plant breeding activities designed to support conservation, identifying issues that have been raised by recent case studies on maize in Mexico. The evidence reported here cannot necessarily be generalized to other crops, farmers, or regions, for three major reasons. First, some of the issues we raise regarding the benefits from collaborative plant breeding are specific only to outcrossing crop species like maize. Second, farmers' seed selection practices differ for maize because the seed is typically selected based on the characteristics of the entire ear—representing single (maternal) plant selection. Third, the evidence is limited geographically to Mexico, the center of origin and diversity for maize. The history of a species in an area is one factor affecting the structure of its genetic diversity and farmers' practices. Research in secondary centers of diversity may raise distinct issues or lead to different conclusions. Similarly, the unique sociocultural and economic characteristics of Mexico, deriving in part from its history, topography, and location, imply that caution must be exercised in extrapolating findings to other nations.

The research described in this chapter consists of a set of pilot studies funded wholly or partially through CIMMYT, and additional work is underway. Considerable work has already been conducted on this subject by other international research centers, such as the International Center for Tropical Agriculture (CIAT), and a review summarizing experiences in international participatory breeding efforts has recently been commissioned. Non-governmental organizations and other institutions are already implementing collaborative plant breeding, but few of these efforts have been documented or published.

The next section presents relevant terms and basic economic concepts used in analyzing the questions of farmers' incentives to pursue collaborative plant breeding. The third section presents the key issues raised and evidence gathered in the case studies. In each study, biological and social factors are intertwined, and failure to consider one or the other set of factors leads to an incomplete assessment of the system of interest. The final section recapitulates conclusions and identifies unresolved issues that need to be addressed in future research.

15.2. DEFINITIONS OF TERMS

15.2.1. Farmers' Management of Diversity

To conceptualize economic issues related to collaborative plant breeding as a policy incentive for genetic resource conservation, we need to understand the process by which farmers select and manage their planting material and their crop populations. Bellon, Pham, and Jackson (1997) have identified three components of farmers' management of diversity. "Variety choice" is the process by which farmers decide which crop varieties to plant. The term "seed flows" (flows of seed or other planting material) refers to the process by which farmers obtain the physical unit of planting material from a given variety that they will grow. The material a farmer plants may have been selected from his or her own crop in the preceding season, exchanged or purchased from other farmers or institutions, or derived from a combination of sources. "Selection and management" is the process by which a farmer who retains planting material from his or her own crop (1) selects the material to be used for planting and (2) handles the material from harvest to planting.

These components can be understood as the dependent or behavioral variables in a collaborative plant breeding activity whose purpose is to support on-farm conservation of crop genetic resources (Bellon and Smale, 1998; Cleveland, Soleri, and Smith, 1998). Although these definitions apply across crop species, in the remainder of this chapter the word "seeds" refers to *maize* planting material. "Farmers' varieties" are defined as crop populations that farmers identify and name as distinct local units, be they landraces, modern varieties, or modern varieties that farmers have selected or mixed with their own landraces. This definition reflects the farmer's own perspective about when a population "becomes" local, irrespective of its biological origin or where he or she procured it (see Soleri and Cleveland, 1993)

For maize in Mexico, and perhaps for other crops and regions, it is important to distinguish a named variety from the physical unit that the farmer plants as seed. Recognizing farmers' practice of introducing varieties (and seed for the same varieties) from the stocks of other farmers, Louette (1994) developed the concept of a "seed lot." A seed lot consists of all kernels of a specific type of maize or variety selected by a farmer and planted during a cropping season to reproduce that particular maize type or variety. A variety is then constituted of all of the seed lots that a number of farmers refer to with the same name. *A seed lot is a physical entity; a variety is associated with a name.* Louette used the seed lot as the unit of analysis for characterizing the intra- and inter-varietal structure of diversity in maize.

15.2.2. Farmers' Incentives to Maintain Diversity

In collaborative plant breeding, the costs of conservation are shifted from the publicly funded, classical system of *ex situ* conservation to farmers themselves, farmers' associations, and other non-governmental organizations. Although the costs of local crop improvement or maintenance are already borne by farmers and their communities, collaborative plant breeding is likely to introduce additional costs. To participate in collaborative plant breeding, individual farmers clearly must perceive the benefits from participation as greater than the costs, including the opportunity cost of the time they devote to it.

Further, if collaborative plant breeding is to provide a link between agricultural development and genetic resource conservation, we must be able to identify participatory strategies for crop improvement that generate both private benefits to farmers and public benefits to society. For farmers who both consume and sell their products, we consider "private benefits" to be the value of the output on the market or in home consumption, including the consumption qualities of grain, fodder, and other non-grain products. Private benefits are as perceived by farmers, and may include either the enhanced market or the non-market value of the crop or its traits, as well as other personal satisfaction derived from cultural aspects of the product. "Public benefits" are the perceived potential genetic gains accruing to society as a whole from future use of crop genetic resources by either farmers or professional plant breeders.

Seed has both private and public attributes. When the farmer chooses to grow a certain amount of seed of a variety or varieties, he or she benefits from the crop's output. That choice also affects the genetic diversity of the crop in the region, measured in observable characteristics and populations as well as the unobservable (or observable only with molecular techniques) alleles of interest to conservationists. Genetic diversity is a public attribute: it is impossible for a farmer to observe or predict the effects of his or her own planting decisions as well as those of the numerous other small-scale farmers in the region on genetic diversity in any given year. The extent to which individual farmers consider the relationship of their variety choices, seed selection, and management practices to those of other farmers in their community is a matter of empirical investigation (Cleveland, Soleri, and Smith, 1998; Cleveland and Murray, 1997; Smale, Bellon and Aguirre, 1998). In any case, the configuration of varieties and the area they cover are determined by farmers' choices in meeting their own objectives and may not necessarily be the most desirable for society from the point of view of genetic diversity (see also chapter 14).

As the previous discussion indicates, if farmers are to cultivate varieties that are defined as socially valuable genetic resources but are not necessarily of private value to them within their farming system, incentives must be provided. How can the objectives of individual farmers and those of society be made more compatible? If farmers perceive private benefits from collaborative plant breeding while it contributes to maintaining the genetic diversity of the crop, both farmers and society as a whole will have gained. In the case reported for the Philippines in this volume (chapter 6), Bellon *et al.* identified a cluster of rice varieties that were both genetically diverse at the molecular level and highly valued by farmers for their consumption characteristics and tolerance to biotic and abiotic stresses. Although still cultivated in the rainfed rice production system, in the irrigated rice production system these varieties had been "discarded" for newer varieties. In the irrigated production system, varieties with shorter duration could be grown and two crops produced; high opportunity costs in terms of output foregone were associated with growing the older varieties, with their longer growing cycles. A breeding intervention to reduce the length of the growing period might enhance the desirability of the older varieties in the irrigated system, although the trade-offs in terms of other traits and the role of the varieties in the irrigated system would need to be investigated.

Figure 15.1 illustrates a similar notion. In a given region, in a given year, for a collection of varieties that are available to farmers and can be grown with varying spatial distributions, two outputs are produced: those from which farmers derive direct private benefits, and the public good, which is crop genetic diversity in the region. The vertical axis, Y, is a production index, which might include grain, fodder, or other output characteristics from which the farmer derives utility. In the simplest case, Y has a single dimension, as in the case of grain yield. The horizontal axis, Z, represents crop genetic diversity. The total number of varieties is fixed in a region in any given year, although farmers can choose among different combinations of them, planted to different areas. The production possibilities frontier represents the maximum amount of private and public goods (Y,Z), that can be produced, given a fixed set of resources. The concavity of the production possibilities frontier results from the fixity of land and genetic resources in any given cropping season and region, as well as the fact that some combinations generate more genetic diversity while others generate more yield. Different frontiers express different cropping seasons and/or different genetic resource bases, some favoring current production and others favoring genetic diversity that is unrelated to ongoing production.

Society as a whole gains utility from crop output and crop genetic diversity. The social indifference curves express preferences over productivity and conservation. Some societies prefer more productivity, which generates current private benefits, and others prefer to conserve genetic diversity for perceived future benefits. The tangency of the social indifference curve with the production possibilities frontier provides the socially optimal allocation of genetic resources. Collaborative plant breeding can link agricultural development to genetic resource conservation if it is possible to identify breeding strategies that augment productivity in terms of the crop characteristics valued by local farmers and augment genetic diversity among the populations grown in the reference region. Accomplishing these twin goals would unambiguously improve social welfare—regardless of the society in which it occurs.

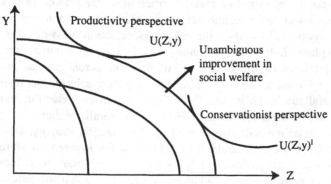

Figure 15.1. Production Possibilities Frontier for Maize Outputs and Genetic Diversity.

15.3. ISSUES RAISED BY RESEARCH ON MAIZE IN MEXICO

Some common features of Mexican farmers' choice of varieties, seed selection, and seed management raise questions about whether they will be able to realize production benefits from their efforts in collaborative breeding. Other features indicate potentially complementary roles of farmers and plant breeders in improving maize landraces on the farm. This section reviews the issues raised by the case studies, which fall into four categories: seed flows; varietal choice; seed selection and management; and farmers' knowledge. To understand these issues fully, however, it is important to understand what is meant by "mass selection" and key biological considerations related to this practice.

15.3.1. Mass Selection in Maize and Implications for Collaborative Plant Breeding

Mexican farmers typically select maize seed based on the ear characteristics of the harvested crop, rather than the characteristics of the plant in the field (SEP, 1982). In Mexico, improved seed selection practices have been recommended, both in the past by the national agricultural research institution and currently by non-governmental organizations. These practices generally include recommendations on selecting the plant from the center of the field in the presence of good competition, followed by the usual selection based on ear characteristics, as well as seed treatment and proper storage (CAECECH 1987; see Rice, Smale, and Blanco, 1997). The practices are intended to improve the effectiveness of farmers' methods of mass selection. With most collaborative plant breeding strategies, some kind of mass selection is likely to form part of the recommendations for maintaining varieties, if not for improving varieties.

"Mass selection" is defined as the identification in a crop population of superior individuals in the form of plants, ears, seed heads, tubers, or stem cuttings, and—in the case of maize—the bulking of seed to form the seed stock for the next generation. If practiced season after season with the same seed stock, mass selection has the potential to maintain or even improve a crop population, depending upon: (1) the extent to which the selected trait is genetically controlled (heritability); (2) genotype–by–environment interaction for the trait; (3) the proportion of the population selected (selection intensity); and (4) gene flow in the form of pollen or seed into the population. Response to mass selection in a cross-pollinated crop like maize is often low. One reason for this low response is that, unless selection for desirable characteristics occurs prior to fertilization and is subsequently controlled based on the selections, selection pressure is exerted only on the maternal plant. Hallauer and Miranda (1988: 213) report that in a composite population, mass selection for grain yield based on only one parent gave an average annual gain of only 1.7%/yr, whereas when both parents were selected the response increased to 7%/yr (Hallauer and Miranda, 1988: 213). Response to selection for traits with greater heritability than yield, such as grain type, is generally higher.

Response to mass selection will also depend on the selection strategy, of which there are many (see Hallauer and Miranda, 1988: 211–215). For low-resource farmers, some form of environmental stratification is likely to be necessary to reduce the confounding effects of environmental variation. In any case, given the parameters involved, an attempt should probably be made to estimate the efficacy of researchers' suggestions to modify local mass selection at the site itself before recommendations are promoted among farmers. In the case of relatively simple qualitative traits such as kernel color, environmental variation will not affect the selection response.

15.3.2. Seed Flows and Seed Lifecycles

The potential of a farmer to reap the rewards of modified mass selection practices also depends on the extent to which he or she retains the seed of distinct populations from harvest to planting, over successive seasons. Louette, Charrier, and Berthaud (1997) have documented that in Cuzalapa Valley of Jalisco, farmers frequently replace, renew, or modify the seed stocks for their varieties by introducing seed obtained from other farmers within and outside the community. Although farmers only rarely pool seed lots of different varieties, poor farmers in particular often mix seed lots considered to be of the same variety to attain enough seed to plant a field. Some farmers believe that they should renew a variety by procuring seed lots for the same variety from other farmers instead of using their own seed year after year.

Table 15.1 shows that for 29 farmers over six cropping cycles, only about half of all of the seed lots, and less than half of the planted area, were from farmers' own harvests. Most of these seed lots were traditional varieties of maize, although some were from advanced generations of modern varieties. The routine use of maize seed stocks produced by other farmers also suggests that in addition to desiring fresh stock, many of these farmers may not have a viable strategy for producing and conserving their own seed.[1]

In southeastern Guanajuato State, which is adjacent to the Bajío (one of the most modernized maize-producing areas of Mexico), Aguirre (1997) also found that some farmers mixed materials in search of "vigor." His research demonstrates, however, that the dominant strategy for procuring seed depends on the different agroecological and economic environments of the farmer (Table 15.2). Aguirre selected his sites to represent contrasts in the degree of market integration and probable length of growing period. Two-thirds of the farmers in the economically isolated environments deliberately introduce and mix their seed. In the more agroecologically marginal and market-integrated environment, the principal strategy is to replace seed. In the most favorable agroecological and economic environment, some farmers retain seed and some replace it. As in the Louette study, most of the maize materials Aguirre found among farmers are traditional varieties or advanced generations of modern varieties.

Table 15.1. Origin of Maize Seed Planted in Cuzalapa, Jalisco, Mexico

Source	Percentage of seed lots	Percentage of area planted
Own seed	52.9	44.9
Seed obtained from other producers in Cuzalapa	35.7	39.9
Seed obtained from producers in other communities	11.4	15.1

Source: Louette (1994).
Note: 29 farmers, 6 cycles, mostly traditional varieties.

Table 15.2. Seed Sources by Environment, Southeastern Guanajuato, Mexico

	80 days moisture		140 days moisture	
Usual practice	Isolated	Integrated	Isolated	Integrated
	% producers			
Save seed from year to year	33	23	20	40
Deliberately introduce or mix seed	67	12	69	14
Replace seed	–	65	11	46

Source: Aguirre (1997).
Note: Most are traditional varieties or advanced generations of modern varieties. 160 farmers. Percentage
    distributions differ significantly by zone with chi-squared test ($\alpha=.01$)

This finding shows that farmers differ with respect to their seed procurement practices, even for traditional varieties. It also suggests that we may need to consider different types of collaborative plant breeding strategies for different types of farmers or environments. For example, modified mass selection techniques may be inappropriate for farmers who routinely replace, introduce, or mix their seed. Supplying those farmers with a greater range of finished or unfinished materials from which they can choose and select may be more beneficial.

Rice, Smale, and Blanco (1997) recorded in detail the use of recommended techniques over five growing seasons by a small group of farmers participating in the initiatives of the Proyecto Sierra de Santa Marta, a non-governmental organization in Veracruz State. The techniques were introduced to encourage farmers to continue growing traditional varieties by improving them. Farmers had identified maize production problems associated with the late maturity and tall stature of their traditional varieties, and they were taught to select for shorter plant height in the field. Approximately 100 farmers in four communities received seed selection and management training, consisting of: (1) marking desirable plants with chalk or a tie; (2) selecting plants within five rows from the boundaries of the field, to reduce the effects of cross-pollination from adjacent fields; (3) selecting plants under good competition with large ears, to ensure healthy, robust plants; (4) after harvest, selecting seed ears from the ears of marked plants based on other desirable ear characteristics; (5) using seed from the center of the cob only; and (6) dusting the maize seed with insecticide or ash and storing it separately from maize grain in a dry place.

In two of the four communities in which workshops were held, only 16 farmers continued to showed interest in the practices several years later. The percentage of seed lots selected from plants declined in each season and seems to have disappeared entirely by the end of the study period. One reason for lack of continuity in the practices was clearly the time cost of labor. Fields were dispersed on steep slopes, farmers had to make separate trips to mark plants, and maize production competed for labor with coffee production at peak periods. Another key factor was undoubtedly the harsh conditions in which these farm families live and produce maize—which led to a high rate of "seed mortality" (Sperling and Loevinsohn, 1983). The Sierra de Santa Marta is an indigenous zone on the edge of a rain-forested volcano, and the maize crop is continually threatened by winds and tropical storms.

Blanco's idea is that the lifecycle of maize seed is closely interwoven with the farmer's life, as depicted for one farmer in Figure 15.2. Many seed introductions "die" several years later, including introductions of traditional and modern varieties, because of external factors such as tropical storms that result in meager harvests, or internal factors such as a change in family structure. When a farmer reports that a variety has been grown for 20 years, he or she may have changed seed lots a number of times during that period.

For new introductions, these patterns may hold whether or not farmers consider the varieties to be promising; for popular varieties, these patterns may hold even when the varieties are considered to be a mainstay of the local cropping system. Similar findings with respect to the loss of seed of traditional and modern varieties, as well as the high rate of change in seed because of renewal, replacement, and hybridization, have been cited by Almekinders, Louwaars, and de Bruijn (1994) for maize and beans in Mesoamerica, and for beans in the Great Lakes Region and eastern Africa (David, 1997; Sperling, Scheideggger, and Buruchara, 1996).

15.3.3. Seed Selection and Storage

In retrospect it seems clear that the practices proposed by the Proyecto Sierra de Santa Marta may have been inappropriate or unsound given conditions at the site. Based on studies conducted under controlled conditions, Hallauer and Miranda (1988: 116) reported an average estimated heritability of 59.6% for plant height. Under the difficult and variable field conditions in the Sierra de Santa Marta, however, heritability would be substantially lower. In other words, well over 40% of the visible variation for plant height would be caused by non-heritable sources of variation, including environmental variation. Since no attempt was made to control environmental variation, the response to selection was probably low.

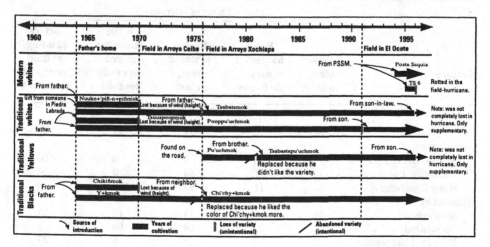

Figure 15.2. Lifecycle of farmer and his maize seed, Sierra de Santa Marta, Veracruz, Mexico.

The apparent simplicity of seed selection and storage practices is also deceptive; as for most agricultural activities in the calendar of subsistence-oriented or semicommercial farm households, gender needs to be recognized when considering the introduction of new techniques. Certainly in the general literature on maize in Mexico, seed selection has been considered the responsibility of farm men (e.g., SEP, 1982). While that is clearly the impression received when men are interviewed, and even when some women are interviewed, observation of households in the Sierra de Santa Marta has led to some different conclusions.

Based on the results of detailed, repeated interviews with men and women from the same household, Rice, Smale, and Blanco (1997) have represented seed selection as an iterative, continuous process that occurs in several stages (Table 15.3). Four of the phases are discrete events: (1) selection of superior plants in the maize field around flowering time, and separation of ears from marked plants at harvest time; (2) selection of a bulk of maize ears when the harvest is brought to the house; (3) a second selection of a bulk of maize ears from the stored maize at some point between harvest and the next season's planting; and (4) selection close to planting time. The first phase was one of the improved seed selection techniques recommended by the Proyecto Sierra de Santa Marta. The second is a setting aside of seed ears from food grain. The third typically happens when stocks are running low following a poor harvest. The fourth is a final revision of the remaining seed stock at planting. A fifth category—putting aside superior ears as women husk maize to prepare food—is distinct from the other phases in that it is more of a continuous process that occurs intermittently from the time of harvest to planting.

Table 15.3. Stages and Features of Seed Selection for Traditional Maize Varieties, Rainy Season, Sierra de Santa Marta, Mexico, 1995

	Phase of selection				
Feature	From plant, in field[a]	From harvested ears, in house	Review of stored ears, in bulk	While removing stored ears daily for food	Before planting, as final review of seed stock
	% of 56 total seed lots				
Time period	14	50	31	59	97
Storage method					
Shelled and bagged[a]	–	28	9	0	100
Unshelled, bagged	–	61	64	0	0
Hung from rafters	–	11	27	100	0
Selected by					
Men only	60	39	18	12	53
Women only	0	22	64	88	21
Both men and women	40	39	18	0	27

Source: Rice, Smale, and Blanco (1997).
[a] Recommended as part of the modified set of practices.

The participation of women was more evident in some aspects of seed selection than in others. Although women in the households surveyed participated in nearly all phases of selection, when asked directly whether they "select" seed (*seleccionar*), they typically answered "no." They answered "yes" when asked if they "set aside" maize for seed (*apartar*). The term "selection" seems to have a very specific meaning related to the introduced practice of selecting superior plants, or to the selection of seed immediately before planting—both of which were almost exclusively accomplished by men. The handling of maize stocks during the year, and the setting aside of good ears during food processing, was almost exclusively the domain of women. When, as is sometimes recommended, ears from plants marked in the field are reviewed at harvest for desirable ear characteristics, and the seed is shelled and bagged with insecticide, the selection process that later occurs inside the household will be sidestepped.

Both the biological and social implications of proposed changes in farmers' practices must be investigated to determine the impact of practices that are introduced with collaborative plant breeding initiatives. The extent to which women are involved in seed selection in other communities, and whether the exclusion of women's seed selection would have a neutral, positive, or negative effect (1) on the genetic structure of the maize population, (2) on the household, or (3) on farm women's well-being, remains to be studied. For example, it is likely that selecting, shelling, and bagging seed with insecticides just after harvest will cause more substantial changes in the populations' genetic structure than would eliminating women's practice of setting aside superior ears when preparing food. Yet many aspects of these selection systems are not understood; it may be that women's handling and observation of the ears contributes information to the formation of the ear ideotype sought by all members of the household. In general, we cannot assume *a priori* that the introduction of modified selection practices makes farmers' selection more "efficient," especially if other selection practices or activities (such as women's) are eliminated in the process of adopting the modified practices.

15.3.4. Choice of Variety

One of the major promises of collaborative breeding is that it may enhance farmers' effectiveness in choosing varieties for characteristics of importance to them. Which varieties should farmers and professional breeders seek to improve? It is well known that farmers in subsistence-oriented and semicommercial agriculture, or farmers who both sell and consume their own crop output, consider more characteristics than just grain yield and grain price when choosing the varieties they grow. Recognition of this fact has led agricultural economists to apply characteristics models (Adesina and Zinnah, 1993) and multi-output models (Renkow and Traxler, 1994) to the analysis of farmers' choice of variety.

Bellon's (1996) treatment of this aspect of farmer behavior focuses on the way in which the vector of characteristics of importance to farmers (which he calls "concerns") changes in dimension with farmers' adaptation to economic change in the environment in which they live. The dimension may expand or contract, provoking changes in the relative importance of characteristics that farmers value. These changes, in turn, lead to the adoption of new varieties and the abandonment of those which are "inferior" under the new set of conditions and constraints.

Varieties tend to be lost when changes in the local biophysical or sociocultural environment reduce the importance of their environmental or sociocultural adaptation (Soleri and Cleveland, 1993). When biophysical or sociocultural changes make it possible to replace varieties, their actual replacement or abandonment appears to be determined by the availability of seed of new varieties that are similar to existing varieties in their growing characteristics, or the supply of alternatives to products made with present varieties.

The interaction between these factors in determining the fate of a particular variety may be complex, as illustrated in the case of Hopi maize folk varieties and vegetables grown by Hopi gardeners. Hopis retain their blue maize varieties because they are adapted to drought and a short growing season, and meet cultural requirements (blue maize is important in religious ceremonies). However, data suggest that the three or more Hopi varieties of blue maize are being collapsed into one, apparently for a number of reasons. One reason may be that today many farmers have full-time jobs in addition to farming; they do not have the time to maintain many different varieties by sowing populations separately to control cross-pollination and then selecting and storing each set of seed. Another reason may be that the importance of the varieties' unique grain characteristics has been diminished by social changes. For example, the introduction of machine grinding reduced the importance of the softer blue maize variety, while the cash economy reduced the desirability of the better storage qualities of the harder blue maize varieties, since storing two years' harvest against crop failure is no longer necessary (Soleri and Cleveland, 1993). We need to understand farmers' adaptation to change and the factors which determine their choice of varieties to be able to predict, for policy purposes, which populations are most likely to be grown and which households are more likely to grow them (Bellon and Smale, 1998; chapters in part III).

15.3.5. Farmers' Knowledge

Comprehending farmers' subjective understanding of their crop genetic resources is important because it shapes their behavior and affects their crop varieties and farming systems in ways that can be measured objectively. It seems likely as well that understanding farmers' perceptions may contribute to a more viable collaborative plant improvement effort.

Louette's findings from Cuzalapa indicate that farmers' seed selection practices protect the phenological integrity of their traditional maize varieties as they define them, despite numerous factors contributing to genetic instability (Louette and Smale, 1998). Analysis of morphological and molecular (isozyme) data suggests that when farmers' varieties are subjected to significant gene flow through cross-pollination, ear characteristics and linked traits are maintained through farmers' selection even though other characteristics may continue to evolve genetically. These findings indicate that there may be further scope for varietal improvement and potentially complementary roles for professional breeders and Mexican farmers in developing methods to improve maize landraces on the farm.

However, there are indications that farmers' own expectations of what they can achieve through seed selection are limited. Repeated, informal interviews with a small sample of farmers in Cuzalapa suggest that these farmers do not see seed selection as a tool for modifying or improving their varieties (Table 15.4). According to them, seed selection is a means of assuring production but not of transforming a variety. Seed selection also protects the "legitimacy" of a variety (in their words, selection is done "*para que salga legítimo*"). Farmers would change from one variety to another before attempting to change a variety through seed selection.

This finding in no way suggests that farmers are "backward" in their thinking. Instead, it suggests that they may know very well what can be accomplished on their farms with the methods currently available to them. In Cuzalapa, few farmers may produce maize in an environment in which they will be able to obtain perceptible benefits from mass selection within the time horizon that they consider relevant to their decisions.

While recognizing the value of documenting farmers' seed procurement and selection practices, Cleveland, Soleri, and Smith (1998) argue that identifying and understanding the genetic perceptions that underlie them will ultimately provide a more versatile and robust tool for collaborative plant breeding. In two communities in the Central Valleys of Oaxaca State, they have used participant observation and formal interviews structured around a series of scenarios to elicit farmers' perceptions of genetic diversity, heritability, selection expectations, and genotype–by–environment interactions. Preliminary results suggest that while farmers may recognize genetic variation in their maize populations, they cannot make use of that variation with the techniques they possess at present, because of the large amount of environmentally caused variation in their fields. Among traits of importance to them, farmers make clear distinctions between those with relatively high and low heritabilities and respond accordingly in their selection efforts, and some farmers recognize segregation among offspring of a single parental phenotype. Farmers in these locations stated that maintaining different maize varieties for different field locations or types does not warrant the effort. Researchers have interpreted this response as an indication that environmental variation within fields was likely to be greater than variation between fields.

Cleveland, Soleri, and Smith have pointed out that their findings describe farmers' perceptions and behavior in terms accessible to outside scientists, allowing those

Table 15.4. Farmers' Perceptions about Seed Selection and Its Purpose, Cuzalapa, 1997 (in order of decreasing frequency; 25 farmers)

Question	Most frequent response
Which ears do you select?	Well-developed, well-filled ears Large ears
Why do you select seed?	To ensure germination To reproduce the variety as we know it
Can you change the characteristics of a variety?[a]	By changing planting dates, using fertilizer, or planting it close to another variety, but *not by seed selection*

Source: Louette and Smale (1998).

[a] Two types of change were discussed: length of growing period (plant characteristic) and number of rows (ear characteristics).

scientists to use their own knowledge and skills more effectively in collaboration with farmers. Farmers' knowledge is not being tested in this work, nor is it being compared against a "correct" or "scientific" template, as many other factors that lie beyond the narrow parameters of the research accomplished to date contribute to farmers' knowledge about their crops.

15.4. CONCLUSIONS AND UNRESOLVED ISSUES

Collaborative or participatory plant breeding uses the skills and experience of both farmer-breeders and professional plant breeders to improve crop plants. The extent of participation by farmer- and professional breeders varies by case and includes, for example, the identification of characteristics for improvement, choice of varieties, and revision of seed selection practices. One of the proposed goals of collaborative plant breeding is to support on-farm conservation. Proponents of this approach argue that while professional plant breeders have conventionally sought to develop fewer varieties adapted to a wider range of environments, participatory breeding can support the maintenance of more diverse, locally adapted plant populations. Both the biological validity and economic feasibility of this proposition require testing. In this chapter, we have used evidence from case studies of maize farming in Mexico to highlight certain key issues that affect farmers' incentives to engage in such efforts. In each case, biological and social factors are interrelated.

The assumption that modified mass selection practices will benefit farmers cannot be generalized and requires technical investigation—particularly given the patterns of seed exchange and seed mortality mentioned in this chapter, but even for farmers who are able to retain their own seed from harvest to planting, year after year. The effect of farmers' use of either improved or traditional mass selection practices to enhance maize yield or other characteristics *under their own conditions* is not well understood. Modest responses to selection in variable environments, combined with the added costs incurred by collaborative plant breeding, may provide weak incentives for farmers to continue growing crop populations identified as important genetic resources, especially compared to the force of the economic changes they face.

Other issues also remained unresolved. To develop collaborative plant breeding strategies that will have an impact, we need to understand how farmers differ with respect to their management of seed and varieties. Decisions will need to be made. Which farmers should outsiders work with? Which outsiders should farmers work with? To assure that the efforts will be sustained on a community basis, we also need to know in what ways germplasm and practices move from farmer to farmer. We need to understand the "social infrastructure" of the exchange of seed and knowledge among farmers (see Ashby *et al.* 1996).

Asking non-governmental organizations and farmers themselves to improve and diffuse varieties does not make genetic resource conservation cheaper—it only shifts the cost burden from some members of society to others. The question remains if the benefits to be achieved will outweigh those new costs. Collaborative plant breeding promises to benefit farmers who have never benefited from the diffusion of modern varieties through formal seed systems, either because the new seed is not adapted to local agronomic conditions, or local preferences favor certain consumption characteristics, or there are few incentives for the development of commercial seed systems in their locality. Depending on the collaborative breeding strategy that is chosen, it also promises heavy time costs for farmers. Participatory research typically requires a lot of local institutional support and a strong cultural basis, as well as a portfolio of income-earning activities, to involve farmers on a long-term basis.

There is a indeed a keen irony in the notion that some of the world's poorest, most neglected farmers are being asked to shoulder the burden of genetic resource conservation for the rest of society and the world. Aside from collaborative plant breeding, other incentives may be provided in the form of subsidies for producers of landraces in selected regions, although most believe that direct payments would not be advisable from an administrative standpoint. If farmers were paid a premium to grow a particular variety, all would choose to grow it—defeating the purpose altogether. The development of niche markets, seed exchanges, and educational campaigns have also been proposed as alternatives or may be considered in combination.

How big an advantage must be generated by the collaborative plant breeding initiative? The evidence presented here raises questions about the magnitude of the benefits maize farmers in Mexico can obtain through participatory plant breeding. Benefits of the size obtained by initial adopters of green revolution crop varieties are hard to envisage. However, the purpose of collaborative breeding strategies is not to replicate the technical changes of the green revolution but to reach farmers who might never have benefited from crop improvement research without participatory initiatives.

There is certainly far more that is unknown than is understood about collaborative plant breeding and its impact on local crop improvement and genetic resource conservation. It seems clear that the approach has potential for providing benefits in all three of those areas. Still, assessing that potential and entering a collaboration that holds the most promise for a positive outcome will require careful multidisciplinary research as well as an examination of researchers' assumptions about the sociocultural, economic, and biological aspects of traditionally based crop selection and management.

Acknowledgments

The authors thank the farmers who have collaborated with them in communities of the States of Oaxaca, Jalisco, Guanajuato, and Veracruz, Mexico. The research conducted by the authors and farmers was funded or supported by a large number of institutions, including: the Association of Women in Science; the Fulbright-Mexico Commission; the Government of France; the International Maize and Wheat Improvement Center; the Rockefeller Foundation; the Proyecto Sierra de Santa Marta; the Committee on Research, Academic Senate; University of California, Santa Barbara; University of Guadalajara; Universidad Nacional Autónoma de México; and the United States Agency for International Development.

Notes

1 In interpreting these results, it may be important to recognize that the community of Cuzalapa is located in the buffer zone of a the Biosphere of the Sierra de Manantlán, Jalisco. Because of the area's status as a buffer zone, research may be undertaken in Cuzalapa but only farmers themselves may introduce seed. Farmers in that community share traditional cultural practices and live in a relatively isolated geographical area, although they are affected by numerous modern and external factors, including labor migration and changes in road infrastructure.

References

Adesina, A. A., and M. N. Zinnah. 1993. Technology characteristics, farmer perceptions and adoption decisions: A tobit model application in Sierra Leone. *Agricultural Economics* 9: 297–311.

Aguirre, A. 1997. *Análisis regional de la diversidad del maíz en el Sureste de Guanajuato*. Ph.D. thesis, Universidad Nacional Autónoma de México, Facultad de Ciencias, Mexico City.

Almekinders, C. J. M., N. P. Louwaars, and G. H. de Bruijn. 1994. Local seed systems and their importance for an improved seed supply in developing countries. *Euphytica* 78: 207–216.

Ashby, J.A., T. Gracia, M. del Pilar Guerrero, C.A. Quirós, J.I. Roa, and J.A. Beltrán. 1996. Innovation in the organization of participatory plant breeding. In Eyzaguirre, P. and M. Iwanaga (eds.), *Participatory Plant Breeding. Proceedings of a Workshop on Participatory Plant Breeding 26–29 July 1995, Wageningen, the Netherlands*. Rome: International Plant Genetic Resources Institute (IPGRI).

Biggs, S. D. 1989. *Resource-Poor Farmer Participation in Research: A Synthesis of Experiences from Nine National Agricultural Research Systems*. OFCOR-Comparative Study Paper No. 3. The Hague: International Serve for National Agricultural Research (ISNAR).

Bellon, M. R. 1996. The dynamics of crop infraspecific diversity: A conceptual framework at the farmer level. *Economic Botany* 50: 26–39.

Bellon, M. R. and S. B. Brush. 1994. Keepers of maize in Chiapas, Mexico. *Economic Botany* 48: 196–209.

Bellon, M. R., J. L. Pham, and M. T. Jackson. 1997. Genetic conservation: A role for rice farmers. In N. Maxted, B.V. Ford-Lloyd, and J.G. Hawkes (eds.), *Plant Genetic Conservation: The In Situ Approach*. London: Chapman and Hall.

Bellon, M. R. and M. Smale. 1998. *A Conceptual Framework for Valuing On-Farm Genetic Resources. CIMMYT Economics Working Paper*. Mexico, D.F.: International Maize and Wheat Improvement Center (CIMMYT).

Berg, T. 1995. Devolution of plant breeding. In L. Sperling and M. Loevinsohn (eds.), *Using Diversity: Enhancing and Maintaining Genetic Resources On-Farm. Proceedings of a Workshop Held on 19–21 June, 1995, in New Delhi*. New Delhi: International Development Research Centre (IDRC).

CAECECH (Campo Agrícola Experimental Centro de Chiapas). 1987. *Guía para la Asistencia Técnica Agrícola: Area de Influencia del Campo Experimental*. Ocoxocoautla de Espinosa, Chiapas: Centro de Investigaciones Forestales y Agropecuarias en el Estado de Chiapas.

Ceccarelli, S., E. Bailey, S. Grando, and R. Tutwiler. 1997. Decentralized, participatory plant breeding: A link between formal plant breeding and small farmers. In CGIAR System-wide Project (ed.), *New Frontiers in Participatory Research and Gender Analysis: Proceedings of the International Seminar on Participatory Research and Gender Analysis for Technology Development*. Cali: Consultative Group on International Agricultural Research (CGIAR) System-Wide Project.

Cleveland, D. A., and S. C. Murray. 1997. The world's crop genetic resources and the rights of indigenous farmers. *Current Anthropology* 38: 477–515.

Cleveland, D. A., D. Soleri, and S. E. Smith. 1994. Do folk crop varieties have a role in sustainable agriculture? *BioScience* 44:740–51.

Cleveland, D. A., D. Soleri, and S. E. Smith. 1998. Farmer varietal management and plant breeding from a biological and sociocultural perspective: Implications for collaborative breeding. Draft Economics Working Paper. Mexico, D.F.: International Maize and Wheat Improvement Center (CIMMYT).

David, S. 1997. *Dissemination and Adoption of New Technology: A Review of Experiences in Bean Research in Eastern and Central Africa, 1991–1996*. Occasional Publication Series, No. 21. Kampala: Network on Bean Research in Africa, International Center for Tropical Agriculture (CIAT).

Eyzaguirre, P. and M. Iwanaga (eds.). 1996. *Participatory Plant Breeding. Proceedings of a Workshop on Participatory Plant Breeding 26–29 July 1995, Wageningen, the Netherlands*. Rome: International Plant Genetic Resources Institute (IPGRI).

Hallauer, A.R., and J.B. Miranda, 1988. *Quantitative Genetics in Maize Breeding*. Second edition. Ames: Iowa State University Press.

Louette, D. 1994. *Gestion traditionnelle de variétés de maïs dans la réserve de la Biosphère Sierra de Manantlán (RBSM, états de Jalisco et Colima, Méxique) et conservation in situ des ressources génétiques de plantes cultivées*. Ph.D. thesis, École Nationale Supérieure Agronomique de Montpellier, France.

Louette, D., A. Charrier, and J. Berthaud. 1997. *In situ* conservation of maize in Mexico: Genetic diversity and maize seed management in a traditional community. *Economic Botany* 51: 20–38.

Louette, D., and M. Smale. 1998. *Farmers' Seed Selection Practices and Maize Variety Characteristics in a Traditionally-Based Mexican Community*. CIMMYT Economics Working Paper 98-04. Mexico, D.F.: International Maize and Wheat Improvement Center (CIMMYT).

Qualset, C. O., A. B. Damania, A. C. A. Zanatta, and S. B. Brush. 1997. Locally based crop plant conservation. In N. Maxted, B. V. Ford-Lloyd and J. G. Hawkes (eds.), *Plant Genetic Conservation: The In Situ Approach*. London: Chapman and Hall.

Renkow, M., and G. Traxler. 1994. Incomplete adoption of modern cereal varieties: The role of grain-fodder tradeoffs. Selected paper, Annual Meetings of the American Agricultural Economics Association, 7–10 August, San Diego, California.

Rice, E., M. Smale, and J.-L. Blanco. 1998. *Farmers' Use of Improved Seed Selection Practices in Mexican Maize: Evidence and Issues from the Sierra de Santa Marta*. CIMMYT Economics Working Paper 97-03. Mexico, D.F.: International Maize and Wheat Improvement Center (CIMMYT).

SEP (Secretaría de Educación Pública). 1982. *Nuestro maíz-Treinta monografías populares*. Tomo 1 y 2 Mexico, D.F.: Consejo Nacional de Fomento Educativo, Secretaría de Educación Pública.

Smale, M., M. Bellon, and A. Aguirre. 1998. Variety characteristics and the land allocation decisions of farmers in a center of maize diversity. Selected paper, Annual Meetings of the American Association of Agricultural Economists, 2–5 August, Salt Lake City, Utah.

Soleri, D., and D. A. Cleveland. 1993. Hopi crop diversity and change. *Journal of Ethnobiology* 13: 203–31.

Soleri, D., and S. E. Smith. 1995. Morphological and phenological comparisons of two Hopi maize varieties conserved *in situ* and *ex situ*. *Economic Botany* 49: 56–77.

Sperling, L., and M. E. Loevinsohn. 1993. The dynamics of adoption: distribution and mortality of bean varieties among small farmers in Rwanda. *Agricultural Systems* 41: 441–453.

Sperling, L., U. Scheidegger, and R. Buruchara. 1996. *Designing Seed Systems with Small Farmers; Principles Derived from Bean Research in the Great Lakes Region of Africa*. Agricultural Research and Extension Network Paper No. 60. London: Overseas Development Agency.

van Oosterom, E. J., M. L. Whitaker, and E. Weltzien R. 1996. Integrating genotype by environment interaction analysis, characterization of drought patterns, and farmer preferences to identify adaptive plant traits for pearl millet. In M. Cooper and G. L. Hammer (eds.), *Plant Adaptation and Crop Improvement*. Wallingford: CAB International.

Weltzien R. E., M. L. Whitaker, and M. M. Anders. 1996. Farmer participating in pearl millet breeding for marginal environments. In P. Eyzaguirre and M. Iwanaga (ed.), *Participatory Plant Breeding: Proceedings of a Workshop on Participatory Plant Breeding, 26–29 July 1995, Wageningen, Netherlands*. Rome: International Plant Genetic Resources Institute (IPGRI).

Witcombe, J. 1997. Decentralization versus farmer participation in plant breeding: Some methodological issues. In CGIAR System-wide Project (ed.), *New Frontiers in Participatory Research and Gender Analysis: Proceedings of the International Seminar on Participatory Research and Gender Analysis for Technology Development*. Cali: Consultative Group on International Agricultural Research (CGIAR) System-Wide Project.

Witcombe, J.R. and A. Joshi. 1995. The impact of farmer participatory research on biodiversity of crops. In L. Sperling and M. Loevinsohn (eds.), *Using Diversity: Enhancing and Maintaining Genetic Resources On-Farm. Proceedings of a workshop held on 19–21 June, 1995, in New Delhi*. New Delhi: International Development Research Centre (IDRC).

Witcombe, J. R., and A. Joshi. 1996. Farmer participatory approaches for varietal breeding and selection and linkages to the formal seed sector. In P. Eyzaguirre and M. Iwanaga (ed.), *Participatory Plant Breeding: Proceedings of a Workshop on Participatory Plant Breeding, 26–29 July 1995, Wageningen, Netherlands*. Rome: International Plant Genetic Resources Institute (IPGRI).

INDEX